Pitman Research Notes in Mathematics Series

Main Editors
H. Brezis, Université de Paris
R.G. Douglas, Texas A&M University
A. Jeffrey, University of Newcastle upon Tyne *(Founding Editor)*

Editorial Board
H. Amann, University of Zürich
R. Aris, University of Minnesota
G.I. Barenblatt, University of Cambridge
A. Bensoussan, INRIA, France
P. Bullen, University of British Columbia
S. Donaldson, University of Oxford
R.J. Elliott, University of Alberta
R.P. Gilbert, University of Delaware
D. Jerison, Massachusetts Institute of Technology
K. Kirchgässner, Universität Stuttgart
B. Lawson, State University of New York at Stony Brook
B. Moodie, University of Alberta
S. Mori, Kyoto University
L.E. Payne, Cornell University
G.F. Roach, University of Strathclyde
I. Stakgold, University of Delaware
W.A. Strauss, Brown University
S.J. Taylor, University of Virginia

Submission of proposals for consideration
Suggestions for publication, in the form of outlines and representative samples, are invited by the Editorial Board for assessment. Intending authors should approach one of the main editors or another member of the Editorial Board, citing the relevant AMS subject classifications. Alternatively, outlines may be sent directly to the publisher's offices. Refereeing is by members of the board and other mathematical authorities in the topic concerned, throughout the world.

Preparation of accepted manuscripts
On acceptance of a proposal, the publisher will supply full instructions for the preparation of manuscripts in a form suitable for direct photo-lithographic reproduction. Specially printed grid sheets can be provided and a contribution is offered by the publisher towards the cost of typing. Word processor output, subject to the publisher's approval, is also acceptable.

Illustrations should be prepared by the authors, ready for direct reproduction without further improvement. The use of hand-drawn symbols should be avoided wherever possible, in order to maintain maximum clarity of the text.

The publisher will be pleased to give any guidance necessary during the preparation of a typescript, and will be happy to answer any queries.

Important note
In order to avoid later retyping, intending authors are strongly urged not to begin final preparation of a typescript before receiving the publisher's guidelines. In this way it is hoped to preserve the uniform appearance of the series.

Addison Wesley Longman Ltd
Edinburgh Gate
Harlow, Essex, CM20 2JE
UK
(Telephone (0) 1279 623623)

Titles in this series. A full list is available from the publisher on request.

100 Optimal control of variational inequalities
 V Barbu
101 Partial differential equations and dynamical systems
 W E Fitzgibbon III
102 Approximation of Hilbert space operators Volume II
 C Apostol, L A Fialkow, D A Herrero and D Voiculescu
103 Nondiscrete induction and iterative processes
 V Ptak and F-A Potra
104 Analytic functions – growth aspects
 O P Juneja and G P Kapoor
105 Theory of Tikhonov regularization for Fredholm equations of the first kind
 C W Groetsch
106 Nonlinear partial differential equations and free boundaries. Volume I
 J I Díaz
107 Tight and taut immersions of manifolds
 T E Cecil and P J Ryan
108 A layering method for viscous, incompressible L_p flows occupying R^n
 A Douglis and E B Fabes
109 Nonlinear partial differential equations and their applications: Collège de France Seminar. Volume VI
 H Brezis and J L Lions
110 Finite generalized quadrangles
 S E Payne and J A Thas
111 Advances in nonlinear waves. Volume II
 L Debnath
112 Topics in several complex variables
 E Ramírez de Arellano and D Sundararaman
113 Differential equations, flow invariance and applications
 N H Pavel
114 Geometrical combinatorics
 F C Holroyd and R J Wilson
115 Generators of strongly continuous semigroups
 J A van Casteren
116 Growth of algebras and Gelfand–Kirillov dimension
 G R Krause and T H Lenagan
117 Theory of bases and cones
 P K Kamthan and M Gupta
118 Linear groups and permutations
 A R Camina and E A Whelan
119 General Wiener–Hopf factorization methods
 F-O Speck
120 Free boundary problems: applications and theory. Volume III
 A Bossavit, A Damlamian and M Fremond
121 Free boundary problems: applications and theory. Volume IV
 A Bossavit, A Damlamian and M Fremond
122 Nonlinear partial differential equations and their applications: Collège de France Seminar. Volume VII
 H Brezis and J L Lions
123 Geometric methods in operator algebras
 H Araki and E G Effros
124 Infinite dimensional analysis–stochastic processes
 S Albeverio
125 Ennio de Giorgi Colloquium
 P Krée
126 Almost-periodic functions in abstract spaces
 S Zaidman
127 Nonlinear variational problems
 A Marino, L Modica, S Spagnolo and M Degliovanni
128 Second-order systems of partial differential equations in the plane
 L K Hua, W Lin and C-Q Wu
129 Asymptotics of high-order ordinary differential equations
 R B Paris and A D Wood
130 Stochastic differential equations
 R Wu
131 Differential geometry
 L A Cordero
132 Nonlinear differential equations
 J K Hale and P Martinez-Amores
133 Approximation theory and applications
 S P Singh
134 Near-rings and their links with groups
 J D P Meldrum
135 Estimating eigenvalues with *a posteriori/a priori* inequalities
 J R Kuttler and V G Sigillito
136 Regular semigroups as extensions
 F J Pastijn and M Petrich
137 Representations of rank one Lie groups
 D H Collingwood
138 Fractional calculus
 G F Roach and A C McBride
139 Hamilton's principle in continuum mechanics
 A Bedford
140 Numerical analysis
 D F Griffiths and G A Watson
141 Semigroups, theory and applications. Volume I
 H Brezis, M G Crandall and F Kappel
142 Distribution theorems of L-functions
 D Joyner
143 Recent developments in structured continua
 D De Kee and P Kaloni
144 Functional analysis and two-point differential operators
 J Locker
145 Numerical methods for partial differential equations
 S I Hariharan and T H Moulden
146 Completely bounded maps and dilations
 V I Paulsen
147 Harmonic analysis on the Heisenberg nilpotent Lie group
 W Schempp
148 Contributions to modern calculus of variations
 L Cesari
149 Nonlinear parabolic equations: qualitative properties of solutions
 L Boccardo and A Tesei
150 From local times to global geometry, control and physics
 K D Elworthy

151 A stochastic maximum principle for optimal control of diffusions
  U G Haussmann
152 Semigroups, theory and applications. Volume II
  H Brezis, M G Crandall and F Kappel
153 A general theory of integration in function spaces
  P Muldowney
154 Oakland Conference on partial differential equations and applied mathematics
  L R Bragg and J W Dettman
155 Contributions to nonlinear partial differential equations. Volume II
  J I Díaz and P L Lions
156 Semigroups of linear operators: an introduction
  A C McBride
157 Ordinary and partial differential equations
  B D Sleeman and R J Jarvis
158 Hyperbolic equations
  F Colombini and M K V Murthy
159 Linear topologies on a ring: an overview
  J S Golan
160 Dynamical systems and bifurcation theory
  M I Camacho, M J Pacifico and F Takens
161 Branched coverings and algebraic functions
  M Namba
162 Perturbation bounds for matrix eigenvalues
  R Bhatia
163 Defect minimization in operator equations: theory and applications
  R Reemtsen
164 Multidimensional Brownian excursions and potential theory
  K Burdzy
165 Viscosity solutions and optimal control
  R J Elliott
166 Nonlinear partial differential equations and their applications: Collège de France Seminar. Volume VIII
  H Brezis and J L Lions
167 Theory and applications of inverse problems
  H Haario
168 Energy stability and convection
  G P Galdi and B Straughan
169 Additive groups of rings. Volume II
  S Feigelstock
170 Numerical analysis 1987
  D F Griffiths and G A Watson
171 Surveys of some recent results in operator theory. Volume I
  J B Conway and B B Morrel
172 Amenable Banach algebras
  J-P Pier
173 Pseudo-orbits of contact forms
  A Bahri
174 Poisson algebras and Poisson manifolds
  K H Bhaskara and K Viswanath
175 Maximum principles and eigenvalue problems in partial differential equations
  P W Schaefer
176 Mathematical analysis of nonlinear, dynamic processes
  K U Grusa
177 Cordes' two-parameter spectral representation theory
  D F McGhee and R H Picard
178 Equivariant K-theory for proper actions
  N C Phillips
179 Elliptic operators, topology and asymptotic methods
  J Roe
180 Nonlinear evolution equations
  J K Engelbrecht, V E Fridman and E N Pelinovski
181 Nonlinear partial differential equations and their applications: Collège de France Seminar. Volume IX
  H Brezis and J L Lions
182 Critical points at infinity in some variational problems
  A Bahri
183 Recent developments in hyperbolic equations
  L Cattabriga, F Colombini, M K V Murthy and S Spagnolo
184 Optimization and identification of systems governed by evolution equations on Banach space
  N U Ahmed
185 Free boundary problems: theory and applications. Volume I
  K H Hoffmann and J Sprekels
186 Free boundary problems: theory and applications. Volume II
  K H Hoffmann and J Sprekels
187 An introduction to intersection homology theory
  F Kirwan
188 Derivatives, nuclei and dimensions on the frame of torsion theories
  J S Golan and H Simmons
189 Theory of reproducing kernels and its applications
  S Saitoh
190 Volterra integrodifferential equations in Banach spaces and applications
  G Da Prato and M Iannelli
191 Nest algebras
  K R Davidson
192 Surveys of some recent results in operator theory. Volume II
  J B Conway and B B Morrel
193 Nonlinear variational problems. Volume II
  A Marino and M K V Murthy
194 Stochastic processes with multidimensional parameter
  M E Dozzi
195 Prestressed bodies
  D Iesan
196 Hilbert space approach to some classical transforms
  R H Picard
197 Stochastic calculus in application
  J R Norris
198 Radical theory
  B J Gardner
199 The C^*-algebras of a class of solvable Lie groups
  X Wang
200 Stochastic analysis, path integration and dynamics
  K D Elworthy and J C Zambrini

201 Riemannian geometry and holonomy groups
   S Salamon
202 Strong asymptotics for extremal errors and polynomials associated with Erdös type weights
   D S Lubinsky
203 Optimal control of diffusion processes
   V S Borkar
204 Rings, modules and radicals
   B J Gardner
205 Two-parameter eigenvalue problems in ordinary differential equations
   M Faierman
206 Distributions and analytic functions
   R D Carmichael and D Mitrovic
207 Semicontinuity, relaxation and integral representation in the calculus of variations
   G Buttazzo
208 Recent advances in nonlinear elliptic and parabolic problems
   P Bénilan, M Chipot, L Evans and M Pierre
209 Model completions, ring representations and the topology of the Pierce sheaf
   A Carson
210 Retarded dynamical systems
   G Stepan
211 Function spaces, differential operators and nonlinear analysis
   L Paivarinta
212 Analytic function theory of one complex variable
   C C Yang, Y Komatu and K Niino
213 Elements of stability of visco-elastic fluids
   J Dunwoody
214 Jordan decomposition of generalized vector measures
   K D Schmidt
215 A mathematical analysis of bending of plates with transverse shear deformation
   C Constanda
216 Ordinary and partial differential equations. Volume II
   B D Sleeman and R J Jarvis
217 Hilbert modules over function algebras
   R G Douglas and V I Paulsen
218 Graph colourings
   R Wilson and R Nelson
219 Hardy-type inequalities
   A Kufner and B Opic
220 Nonlinear partial differential equations and their applications: Collège de France Seminar. Volume X
   H Brezis and J L Lions
221 Workshop on dynamical systems
   E Shiels and Z Coelho
222 Geometry and analysis in nonlinear dynamics
   H W Broer and F Takens
223 Fluid dynamical aspects of combustion theory
   M Onofri and A Tesei
224 Approximation of Hilbert space operators. Volume I. 2nd edition
   D Herrero
225 Operator theory: proceedings of the 1988 GPOTS–Wabash conference
   J B Conway and B B Morrel
226 Local cohomology and localization
   J L Bueso Montero, B Torrecillas Jover and A Verschoren
227 Nonlinear waves and dissipative effects
   D Fusco and A Jeffrey
228 Numerical analysis 1989
   D F Griffiths and G A Watson
229 Recent developments in structured continua. Volume II
   D De Kee and P Kaloni
230 Boolean methods in interpolation and approximation
   F J Delvos and W Schempp
231 Further advances in twistor theory. Volume I
   L J Mason and L P Hughston
232 Further advances in twistor theory. Volume II
   L J Mason, L P Hughston and P Z Kobak
233 Geometry in the neighborhood of invariant manifolds of maps and flows and linearization
   U Kirchgraber and K Palmer
234 Quantales and their applications
   K I Rosenthal
235 Integral equations and inverse problems
   V Petkov and R Lazarov
236 Pseudo-differential operators
   S R Simanca
237 A functional analytic approach to statistical experiments
   I M Bomze
238 Quantum mechanics, algebras and distributions
   D Dubin and M Hennings
239 Hamilton flows and evolution semigroups
   J Gzyl
240 Topics in controlled Markov chains
   V S Borkar
241 Invariant manifold theory for hydrodynamic transition
   S Sritharan
242 Lectures on the spectrum of $L^2(\Gamma\backslash G)$
   F L Williams
243 Progress in variational methods in Hamiltonian systems and elliptic equations
   M Girardi, M Matzeu and F Pacella
244 Optimization and nonlinear analysis
   A Ioffe, M Marcus and S Reich
245 Inverse problems and imaging
   G F Roach
246 Semigroup theory with applications to systems and control
   N U Ahmed
247 Periodic-parabolic boundary value problems and positivity
   P Hess
248 Distributions and pseudo-differential operators
   S Zaidman
249 Progress in partial differential equations: the Metz surveys
   M Chipot and J Saint Jean Paulin
250 Differential equations and control theory
   V Barbu

251 Stability of stochastic differential equations with respect to semimartingales
  X Mao
252 Fixed point theory and applications
  J Baillon and M Théra
253 Nonlinear hyperbolic equations and field theory
  M K V Murthy and S Spagnolo
254 Ordinary and partial differential equations. Volume III
  B D Sleeman and R J Jarvis
255 Harmonic maps into homogeneous spaces
  M Black
256 Boundary value and initial value problems in complex analysis: studies in complex analysis and its applications to PDEs 1
  R Kühnau and W Tutschke
257 Geometric function theory and applications of complex analysis in mechanics: studies in complex analysis and its applications to PDEs 2
  R Kühnau and W Tutschke
258 The development of statistics: recent contributions from China
  X R Chen, K T Fang and C C Yang
259 Multiplication of distributions and applications to partial differential equations
  M Oberguggenberger
260 Numerical analysis 1991
  D F Griffiths and G A Watson
261 Schur's algorithm and several applications
  M Bakonyi and T Constantinescu
262 Partial differential equations with complex analysis
  H Begehr and A Jeffrey
263 Partial differential equations with real analysis
  H Begehr and A Jeffrey
264 Solvability and bifurcations of nonlinear equations
  P Drábek
265 Orientational averaging in mechanics of solids
  A Lagzdins, V Tamuzs, G Teters and A Kregers
266 Progress in partial differential equations: elliptic and parabolic problems
  C Bandle, J Bemelmans, M Chipot, M Grüter and J Saint Jean Paulin
267 Progress in partial differential equations: calculus of variations, applications
  C Bandle, J Bemelmans, M Chipot, M Grüter and J Saint Jean Paulin
268 Stochastic partial differential equations and applications
  G Da Prato and L Tubaro
269 Partial differential equations and related subjects
  M Miranda
270 Operator algebras and topology
  W B Arveson, A S Mishchenko, M Putinar, M A Rieffel and S Stratila
271 Operator algebras and operator theory
  W B Arveson, A S Mishchenko, M Putinar, M A Rieffel and S Stratila
272 Ordinary and delay differential equations
  J Wiener and J K Hale
273 Partial differential equations
  J Wiener and J K Hale
274 Mathematical topics in fluid mechanics
  J F Rodrigues and A Sequeira
275 Green functions for second order parabolic integro-differential problems
  M G Garroni and J F Menaldi
276 Riemann waves and their applications
  M W Kalinowski
277 Banach C(K)-modules and operators preserving disjointness
  Y A Abramovich, E L Arenson and A K Kitover
278 Limit algebras: an introduction to subalgebras of C^*-algebras
  S C Power
279 Abstract evolution equations, periodic problems and applications
  D Daners and P Koch Medina
280 Emerging applications in free boundary problems
  J Chadam and H Rasmussen
281 Free boundary problems involving solids
  J Chadam and H Rasmussen
282 Free boundary problems in fluid flow with applications
  J Chadam and H Rasmussen
283 Asymptotic problems in probability theory: stochastic models and diffusions on fractals
  K D Elworthy and N Ikeda
284 Asymptotic problems in probability theory: Wiener functionals and asymptotics
  K D Elworthy and N Ikeda
285 Dynamical systems
  R Bamon, R Labarca, J Lewowicz and J Palis
286 Models of hysteresis
  A Visintin
287 Moments in probability and approximation theory
  G A Anastassiou
288 Mathematical aspects of penetrative convection
  B Straughan
289 Ordinary and partial differential equations. Volume IV
  B D Sleeman and R J Jarvis
290 K-theory for real C^*-algebras
  H Schröder
291 Recent developments in theoretical fluid mechanics
  G P Galdi and J Necas
292 Propagation of a curved shock and nonlinear ray theory
  P Prasad
293 Non-classical elastic solids
  M Ciarletta and D Ieşan
294 Multigrid methods
  J Bramble
295 Entropy and partial differential equations
  W A Day
296 Progress in partial differential equations: the Metz surveys 2
  M Chipot
297 Nonstandard methods in the calculus of variations
  C Tuckey
298 Barrelledness, Baire-like- and (LF)-spaces
  M Kunzinger
299 Nonlinear partial differential equations and their applications. Collège de France Seminar. Volume XI
  H Brezis and J L Lions
300 Introduction to operator theory
  T Yoshino

301 Generalized fractional calculus and applications
   V Kiryakova
302 Nonlinear partial differential equations and their applications. Collège de France Seminar Volume XII
   H Brezis and J L Lions
303 Numerical analysis 1993
   D F Griffiths and G A Watson
304 Topics in abstract differential equations
   S Zaidman
305 Complex analysis and its applications
   C C Yang, G C Wen, K Y Li and Y M Chiang
306 Computational methods for fluid-structure interaction
   J M Crolet and R Ohayon
307 Random geometrically graph directed self-similar multifractals
   L Olsen
308 Progress in theoretical and computational fluid mechanics
   G P Galdi, J Málek and J Necas
309 Variational methods in Lorentzian geometry
   A Masiello
310 Stochastic analysis on infinite dimensional spaces
   H Kunita and H-H Kuo
311 Representations of Lie groups and quantum groups
   V Baldoni and M Picardello
312 Common zeros of polynomials in several variables and higher dimensional quadrature
   Y Xu
313 Extending modules
   N V Dung, D van Huynh, P F Smith and R Wisbauer
314 Progress in partial differential equations: the Metz surveys 3
   M Chipot, J Saint Jean Paulin and I Shafrir
315 Refined large deviation limit theorems
   V Vinogradov
316 Topological vector spaces, algebras and related areas
   A Lau and I Tweddle
317 Integral methods in science and engineering
   C Constanda
318 A method for computing unsteady flows in porous media
   R Raghavan and E Ozkan
319 Asymptotic theories for plates and shells
   R P Gilbert and K Hackl
320 Nonlinear variational problems and partial differential equations
   A Marino and M K V Murthy
321 Topics in abstract differential equations II
   S Zaidman
322 Diffraction by wedges
   B Budaev
323 Free boundary problems: theory and applications
   J I Diaz, M A Herrero, A Liñan and J L Vazquez
324 Recent developments in evolution equations
   A C McBride and G F Roach
325 Elliptic and parabolic problems: Pont-à-Mousson 1994
   C Bandle, J Bemelmans, M Chipot, J Saint Jean Paulin and I Shafrir
326 Calculus of variations, applications and computations: Pont-à-Mousson 1994
   C Bandle, J Bemelmans, M Chipot, J Saint Jean Paulin and I Shafrir
327 Conjugate gradient type methods for ill-posed problems
   M Hanke
328 A survey of preconditioned iterative methods
   A M Bruaset
329 A generalized Taylor's formula for functions of several variables and certain of its applications
   J-A Riestra
330 Semigroups of operators and spectral theory
   S Kantorovitz
331 Boundary-field equation methods for a class of nonlinear problems
   G N Gatica and G C Hsiao
332 Metrizable barrelled spaces
   J C Ferrando, M López Pellicer and L M Sánchez Ruiz
333 Real and complex singularities
   W L Marar
334 Hyperbolic sets, shadowing and persistence for noninvertible mappings in Banach spaces
   B Lani-Wayda
335 Nonlinear dynamics and pattern formation in the natural environment
   A Doelman and A van Harten
336 Developments in nonstandard mathematics
   N J Cutland, V Neves, F Oliveira and J Sousa-Pinto
337 Topological circle planes and topological quadrangles
   A E Schroth
338 Graph dynamics
   E Prisner
339 Localization and sheaves: a relative point of view
   P Jara, A Verschoren and C Vidal
340 Mathematical problems in semiconductor physics
   P Marcati, P A Markowich and R Natalini
341 Surveying a dynamical system: a study of the Gray–Scott reaction in a two-phase reactor
   K Alhumaizi and R Aris
342 Solution sets of differential equations in abstract spaces
   R Dragoni, J W Macki, P Nistri and P Zecca

Roberto Dragoni
Università di Siena, Italy

Jack W Macki
University of Alberta, Canada

Paolo Nistri and Pietro Zecca
Università di Firenze, Italy

Solution sets of differential operators in abstract spaces

Addison Wesley Longman Limited
Edinburgh Gate, Harlow
Essex CM20 2JE, England
and Associated Companies throughout the world.

*Published in the United States of America
by Addison Wesley Longman Inc.*

© Addison Wesley Longman Limited 1996

All rights reserved; no part of this publication may be reproduced, stored
in a retrieval system, or transmitted in any form or by any means,
electronic, mechanical, photocopying, recording, or otherwise, without
the prior written permission of the Publishers, or a licence permitting
restricted copying in the United Kingdom issued by the Copyright
Licensing Agency Ltd, 90 Tottenham Court Road, London, W1P 9HE

First published 1996

AMS Subject Classifications: (Main) 34G99, 34-02
                               (Subsidiary) 34A60

ISSN 0269-3674

ISBN 0 582 29450 9

British Library Cataloguing in Publication Data

A catalogue record for this book is
available from the British Library

Printed and bound in Great Britain
by Biddles Ltd, Guildford and King's Lynn

Contents

Definitions and Preliminary Results 1

Peano's Theorem in Infinite Dimensional Spaces 6

Kneser Type Theorems 20

Aronszajn Type Theorems 43

Differential Inclusions in Banach Spaces 59

Boundary Value Problems 74

Bibliography 88

Index of Notation

Symbol	Meaning
S, $S(x_0)$	set of solutions to $\dot{x} = f(t,x)$, $x(t_0) = x_0$
$C([a,b], X)$	continuous functions from $[a,b] \subset \mathbb{R}$ to the linear topological space X
X^*	dual of X
$\sigma(X, X^*)$	weak topology on X
$\sigma(X^*, X)$	weak star topology on X^*
$\lambda(X^*, X)$	topology of precompact convergence on X^*
$\beta(X^*, X)$	strong topology on X^* induced by X
$\mathcal{P}(X)$	power set of X
$\mathcal{P}_K(X)$	compact subsets of X
$\mathcal{P}_{KC}(X)$	compact closed subsets of X
Id	Identity map
$L^1([a,b])$	Lebesgue integrable functions from $[a,b] \subset \mathbb{R}$ to $\mathbb{R}$
$\mu(A)$	Lebesgue measure of A
$L^1(A, X)$	Bochner integrable functions from the measure space A to the linear topological space X
co A	convex hull of A
$\overline{A}$	closure of A

A^0	polar of $A = \{f \in X^* \mid |f(x)| \leq 1 \text{ for } x \in A\}$
$d(x,y)$	distance between points x and y in X
$\tilde{h}(x,A)$	distance between the point x and the set A in $X = \inf_{y \in A} d(x,y)$
$h(U,V)$	Hausdorff distance between sets U and V
$B(x_0, r)$	Ball of radius r centred at x_0
$B(A, \varepsilon)$	$\{x \mid \tilde{h}(x,A) < \varepsilon\}$
$\alpha(V)$	Kuratowski measure of noncompactness (mnc) of V
$\chi(V)$	Hausdorff mnc of V
$\beta_d(V)$	de Blasi mnc of V
$c_A(t)$	characteristic function of the set A
$Gr(F)$	graph of the multifunction $F \equiv \bigcup_{x \in X} \bigcup_{y \in F(x)} (x,y)$

Acknowledgements

The authors gratefully acknowledge the continuing support of the Consiglio Nazionale delle Ricerche d'Italia and the Natural Sciences and Engineering Research Council of Canada.

We express our sincere thanks to Professor Wieslaw Krawcewicz for carefully and critically reading through the manuscript.

Introduction

Let $f(t,x)$ be a continuous function from $[t_0, t_0 + a] \times \mathbb{R}^n$ into $\mathbb{R}^n$ and $x_0 \in \mathbb{R}^n$. Consider the Cauchy problem:

$$\begin{cases} \dot{x}(t) = f(t, x(t)) \\ x(t_0) = x_0. \end{cases} \qquad \text{C}_n)$$

It is well known that the problem $\text{C}_n)$ admits a unique solution locally when f is continuous in (t, x) and Lipschitz in x.

In 1890 Peano (see [110] and [111]) showed that, under the single assumption that f is continuous, the problem $\text{C}_1)$ defined on $\mathbb{R}$ has local solutions, but in general the property of uniqueness does not hold. For x_0 fixed, the set of these solutions is the so-called "Peano funnel."

From this result arose the problem of studying the structure of this set with the aim of describing completely its topological and algebraic properties.

Peano himself showed that if S is the funnel of solutions of the problem $\text{C}_1)$, then the set $S(t) = \{x(t) \colon x \in S\}$, called a section of S, is always nonempty, connected and compact (that is, a continuum) in the standard topology of the real line, for t in a neighborhood of t_0. This result was extended in 1923 by Kneser [78] to differential equations in $\mathbb{R}^n$, and five years later Hukuhara [68] proved that the set S is a continuum in the Banach space of continuous functions with the sup norm.

A more precise characterization of S was found in 1942, when Aronszajn [5] showed that the Peano funnel is homeomorphic to the intersection of a decreasing sequence of compact contractible sets; this implies that S is an acyclic set, i.e. from the point view of cohomology theory S is equivalent to a single point. As Aronszajn said, the proof of his theorem is based on the following idea: if "we can approximate the problem $\text{C}_n)$ with a more regular differential equation admitting a unique solution," then the solution set of $\text{C}_n)$ is an R_δ set (see page 2 for the definition of an R_δ set).

Since then, results proved with this technique are called "Aronszajn type theorems," while results showing that the Peano funnel is a continuum are called "Kneser type theorems." The Aronszajn type of result generally utilizes fixed point theorems for (noncompact) operators in Banach spaces; the Kneser type is obtained by means of a separate proof of nonemptiness, compactness and connectedness, see for example Stampacchia [129] and Yorke [157]. Some important generalizations of Aronszajn's theorem are due, in chronological order since 1967, to Browder and Gupta [17], Vidossich ([151] and [152]) and Lasry and Robert [88]. As an example, here is a result of Aronszajn type, obtained very recently by Bielawski and Pruszko [12]. We recall that $f : [a,b] \times \mathbb{R}^n \to \mathbb{R}^n$ is a Carathéodory function if $f(t,x)$ is measurable in t for each fixed x, continuous in x for each fixed t, and $|f(t,x)| \leq m(t)$ with $m \in L^1(a,b)$, for all (t,x) in a ball containing the point (t_0, x_0).

Theorem. *Let $f : [a,b] \times \mathbb{R}^n \to \mathbb{R}^n$ be a Carathéodory function satisfying $\|f(t,x)\| \leq \alpha(t) + \beta(t)\|x\|$ where $t \in [a,b]$, $x \in \mathbb{R}^n$ and $\alpha, \beta : [a,b] \to [0,\infty)$ are integrable functions.*

Then the solution set of C_n) is an R_δ set in $C([a,b], \mathbb{R}^n)$.

The generalization described above resolves more or less completely the problem of characterizing the Peano funnel of the problem C_n).

We mention the recent work of V.V. Filippov and collaborators [52], which focusses mainly on continuous dependence on parameters but also includes results on the Peano funnel. Given two topological spaces X and Y, he considers the space (A, f) of "partial mappings" $f : A \subset X \to Y$. By proving very general theorems on the space of graphs of these mappings, he can establish for example continuous dependence on parameters for very pathological equations. So far the applications are in finite-dimensional situations, but the results should extend to infinite-dimensional problems.

Z. Artstein [6] has presented sufficient and "as close to necessary as is possible" conditions for continuous dependence on parameters for (not necessarily unique) solutions of operator equations. These results contain information on the structure of the solution set. Heunis [63] has used Artstein's point of view to develop the best possible such theorems for ordinary differential equations in $\mathbb{R}^n$.

Our aim will be to analyze the structure of the solution set of a differential equation in the more general case of abstract topological (infinite dimensional) *spaces.* We will discuss both the spaces with a paucity of useful properties, e.g. locally convex spaces, and the more richly endowed spaces, e.g. Banach or Hilbert spaces; the results obtained will of course be true also in finite dimensional spaces. For single-valued problems we shall always assume that f is continuous or of Carathéodory type, while for multivalued differential equations we shall assume that the map is at least upper semi-continuous (u.s.c.).

Our presentation is organized as follows:

- In the first chapter we present notation, definitions, and preliminary results that will be used throughout.

- Since we want to describe the properties of the solution set, we must establish existence theorems in infinite dimensional spaces, for example we must examine the validity of Peano's existence theorem in abstract spaces. The second chapter deals with this question, in fact the first part contains examples which show that Peano's theorem fails. The second part deals with the few cases of abstract spaces in which, for particular topologies, the hypothesis of continuity of f is sufficient to assure the existence of solutions of C). Finally, we give an unusual characterization of reflexive Banach spaces linked to Peano's theorem.

- In the third chapter we look for further conditions to add to the hypothesis of continuity in order to guarantee the existence of solutions of the Cauchy problem, and we show that under our assumptions the corresponding Peano funnel is a continuum of $C(J, X)$. The results are first stated in Banach spaces and then extended to locally convex spaces. We also study the structure of the solution set for spaces endowed with topologies different from the usual strong topology (for example, the so-called "weak solutions").

- In the fourth chapter we study R_δ sets, which are suitable for a more precise description of the properties of the Peano funnel for the problem C). Some results of Aronszajn type are stated for infinite dimensional Banach spaces.

- The fifth chapter is devoted to the multivalued Cauchy problem. The research in this field is only ten years old and it is very far from being completed. The few known results concerning the solution set are presented.

- The sixth chapter deals with boundary value problems and in particular the Sturm-Liouville problem. The studies on this subject have only recently begun, in particular nothing has yet been proven on the Peano funnel. We shall prove a simple, but new, result on the structure of the solution set of a boundary value problem.

- Finally, we present an extensive bibliography. Since at present there does not exist an exhaustive bibliography on this subject, we hope to at least partly fill this gap.

1. Definitions and Preliminary Results

In this chapter we introduce the definitions which will be used and we state some basic results. We first introduce some topological concepts.

Let X be a linear topological space. A set $A \subset X$ is **balanced** if $\lambda A \subset A$ for $|\lambda| \leq 1$. The set A is called **absorbing** if $\forall\, x \in X\ \exists\, \lambda_0$ such that $x \in \lambda A$ for $|\lambda| \geq \lambda_0$. Given $A,\, B \subset X$, we shall say that A **absorbs** B if $\exists\, \lambda_0$ such that $B \subset \lambda A$ for $|\lambda| \geq \lambda_0$. A set A is **bounded** if it is absorbed by every neighborhood of zero.

An absorbing, balanced, closed and convex subset of X is **a barrel**.

A locally convex linear topological space is **barreled** if each barrel in X is a neighborhood of zero.

By the **convex** (resp. **balanced**) **hull** of a set A we mean the intersection of all convex (resp. balanced) sets which contain A, denoted $\operatorname{co}(A)$. The closure of a set A is denoted $\overline{A}$. The set of all possible linear combinations of the elements of A is called the **linear hull** of A.

If $A \subset X$, the **polar set** of A is $A^0 = \{f \in X^* : |f(x)| \leq 1\ \ \forall\, x \in A\}$.

A linear topological space X is **quasi-complete** if every closed and bounded subset of X is complete.

A Banach space X is called a **weakly compactly generated space** (wcg space) if there exists some weakly compact subset K whose linear hull is equal to X.

A **Schauder basis** of a Banach space X consists of a sequence $\{x_i\}_{i \in N} \subset X$ such that for every point $x \in X$ there exists a unique sequence $\{a_i\}_{i \in N}$ of real numbers such that $\lim\limits_{n \to \infty} \|\sum\limits_{i=1}^{n} a_i x_i - x\| = 0$.

Let X be a topological linear space and $x,\, y \in X$. A **path** from x to y is a continuous function $f : [0,1] \to X$ such that $f(0) = x$ and $f(1) = y$.

A subset $A \subset X$ is said to be **arcwise connected** if for every pair of points $x,\, y \in A$ there exists a path $f : I \to X$ contained in A which joins x to y.

Let A be a subset of a metric space X. A is called a **retract** if there exists a continuous mapping (retraction) $r : X \to A$ such that $r|_A$ coincides with identity map.

A closed subset A of a metric space X is said to be an **absolute retract** (AR) if every homeomorphic image of A in any metric space Y is a retract of Y.

A subset A of a metric space X is **contractible** if there exists a continuous function $h: [0,1] \times A \to X$ (homotopy) and a point $a \in A$ such that $\forall\, x \in A$ we have:

$$h(0, x) = a \quad \text{and} \quad h(1, x) = x.$$

A set $B \subset X$ is **simply connected** if every closed path is contractible to a single point. Recall that the notions of simple and arcwise connectedness are independent. Notice that a contractible set is both simply connected and arcwise connected.

By an R_δ set we mean a set R in a metric space Y which is homeomorphic to the intersection of a decreasing sequence $\{R_n\}_{n \in N}$ of absolute retracts. If every R_n is compact, we shall say that R is a compact R_δ set.

Hyman's Theorem (see [69]) states that a subset of a metric space is a compact R_δ if and only if it is the intersection of a decreasing sequence of contractible compact metric spaces. Observe that every compact R_δ is a continuum (compact, connected and nonempty), but, in contrast to contractible sets, need not be arcwise connected.

Let us recall some basic concepts concerning **cohomology theory.**

Let H^n be the n-dimensional (contravariant) functor of Alexander's cohomology (see Hu [62]) with real coefficients in a group G (e.g. the set Z of integers). A compact metric space A is called **acyclic** if the following conditions are satisfied:

$$H^0(A) = G, \quad H^n(A) = \{0\} \quad \text{if} \quad n \neq 0.$$

In other words an acyclic set has the same cohomology groups as a single point. Some examples of such sets are R_δ (hence contractible) sets and convex sets.

Let X^* be the dual space of a linear space X. The **weak topology on X** is denoted by $\sigma(X, X^*)$ and it is generated by the following fundamental system of neighborhoods of zero:

$$U_{\varepsilon, \{f_1, \ldots, f_n\}} = \{x \in X : |f_i(x)| < \varepsilon \quad \text{for} \quad i = 1, \ldots, n\}$$

where $\varepsilon > 0$ and $\{f_1, \ldots, f_n\}$ is a finite family of elements in X^*.

The **weak star topology** $\sigma(X^*, X)$ **on X^*** is created $\forall\, \varepsilon > 0$ and $\{x_1, \ldots, x_n\} \subset X$ by the following neighborhoods:

$$U_{\varepsilon, \{x_1, \ldots, x_n\}} = \{f \in X^* : |f(x_i)| < \varepsilon \quad \text{for} \quad i = 1, \ldots, n\}.$$

Finally the **topology of precompact convergence** $\lambda(X^*, X)$ on X^* is defined by the neighborhoods of zero

$$U_{\varepsilon, K} = \{f \in X^* : |f(x)| < \varepsilon \text{ for } \forall x \in K\}$$

where $K \subset X$ is any precompact set.

The Banach-Dieudonné theorem (see [66]) deals with the connection between these last two topologies:

Theorem. *Let X be a metrizable locally convex space. Then the topology $\lambda(X^*, X)$ is the finest topology on X^* which induces on every equicontinuous subset of X^* the same topology as $\sigma(X^*, X)$.*

If X is a Banach space, the Ascoli-Arzelá Theorem gives an important characterization of compactness for subsets of $C(J, X)$, the continuous functions from an interval $J \subset \mathbb{R}$ to X:

A subset U of $C(J, X)$ is compact if and only if it it closed, equicontinuous on J and $\forall t \in J$, $\{x(t) \mid x(\cdot) \in U\}$ is totally bounded.

For any metric space, $B(x_0, r)$ is the ball $\{y \mid d(y, x_0) < r\}$, $B_1 = B(0, 1)$.

Given two subsets U, V of a metric space X, the **Hausdorff distance** between U and V is:

$$h(U, V) = \max \left\{ \sup_{x \in U} \tilde{h}(x, V),\ \sup_{x \in V} \tilde{h}(x, U) \right\}$$

where $\tilde{h}(x, U) = \inf_{y \in U} d(x, y)$ and $\tilde{h}(x, V) = \inf_{y \in V} d(x, y)$. By definition $h(x, \emptyset) = +\infty$. We have:

$$h(U, V) = \inf \{\varepsilon > 0 : U \subset B(V, \varepsilon) \text{ and } V \subset B(U, \varepsilon)\},$$

where $B(A, \varepsilon) = \{x \mid \tilde{h}(x, A) < \varepsilon\}$.

For every bounded set $V \subset X$, the **Kuratowski measure of noncompactness** α is defined as:

$$\alpha(V) = \inf \{d > 0 : V \text{ has a finite cover of sets with diameter } \leq d\}.$$

Note that $\alpha(V) \leq 2 \sup_{x \in V} \|x\|$.

The **Hausdorff measure of noncompactness** χ consists of:

$$\chi(V) = \inf\{d > 0 \colon V \text{ admits a finite cover of balls with radius } \leq d\}.$$

Here are some properties of α and χ (let ν stand for α or χ):

(i) $\nu(V) = 0$ if and only if V is precompact
(ii) $\nu(\overline{co}\, V) = \nu(V)$
(iii) $\nu(\bigcup V_\tau) \geq \max_\tau \nu(V_\tau)$
(iv) $\nu(V_1 + V_2) \leq \nu(V_1) + \nu(V_2)$ and $\nu(\lambda V) = |\lambda|\nu(V)$
(v) $\chi(V) \leq \alpha(V) \leq 2\chi(V)$.
(vi) $\forall\, U \subset X$, $\varepsilon > 0$, $\exists\, \delta > 0$ such that $|\nu(U) - \nu(V)| < \varepsilon\ \forall\, V$ satisfying $h(U,V) < \delta$ (continuity).

Here $V_1 + V_2 = \{x + y|\ x \in V_1,\ y \in V_2\}$, $\lambda V = \{\lambda x|\ x \in V\}$.

For a complete treatment of the theory of measures of noncompactness we refer to Akhmerov et al. [1].

If X is a linear topological space, $[a, b] \subset \mathbb{R}$, and $x(t)\colon [a, b] \to X$, then under reasonable conditions on $x(\cdot)$ and X the Bochner integral $\int_a^b x(t)\,dt$ can be defined as follows. For each x^* in the dual X^* of X, form the evaluation $g : [a, b] \to \mathbb{R}$, $g(t) = \langle x(t), x^*\rangle$, then compute $\int_a^b g(t)\,dt$. This defines a continuous linear functional on X^*, i.e., an element $p \in X^{**}$. If p can be identified with an element of X under the natural imbedding of X into X^{**}, then $p = \int_a^b x(t)\,dt$.

If X is locally convex, $x(\cdot) \in C([a, b], X)$, and $\overline{co}\,\{x([a, b])\}$ is sequentially complete, then $x(\cdot)$ is Bochner integrable on $[a, b]$. The mean-value theorem can be written $\int_a^b x(t)\,dt \in (b - a)\overline{co}\,\{x[a, b]\}$.

If X is a normed linear space, then $x(t)\colon [a, b] \to X$ is Bochner integrable if and only if $x(t)$ is strongly measurable and $\|x(t)\|$ is Lebesgue integrable on $[a, b]$.

Let X be a Banach space, $x_0 \in X$, $T \equiv [0, a]$. A function $f : T \times B(x_0, r) \to X$ is called **weak-weak continuous** if $\forall\,(t, x) \in T \times B(x_0, r)$, and $\forall\, U$ a weak neighborhood of $f(t, x)$, $\exists\, \delta > 0$ and a weak neighborhood V of x such that $\forall\, y \in V \cap B(x_0, r)$ and $\forall\, s \in T$ with $|t - s| < \delta$ we have $f(s, y) \in U$.

A sequence of functions $\{f_n\}_{n \in \mathbb{N}}$ with $f_n : T \to X$ converges **weakly uniformly** to f if $\forall\, g \in X^*$ and $\forall\, \varepsilon > 0\ \exists\, n_0 = n_0(g, \varepsilon)$ such that $\forall\, n \geq n_0$ we have $|g(f_n(t)) - g(f(t))| < \varepsilon\quad \forall\, t \in I$.

Dugundji's extension theorem will be useful:

Theorem. *Let X and Y be two Banach spaces, $D \subset X$ a closed set and $f : D \to Y$ a continuous function. Then there exists a continuous extension $\widehat{f} : X \to Y$ of f such that $\widehat{f}(X) \subset \operatorname{co}(f(D))$.*

Finally, let us recall some basic notions about multifunctions.

Let X, Y be metric spaces and $\mathcal{P}(Y)$ the power set of Y. A **multifunction** is a mapping $F : X \to \mathcal{P}(Y)$ which maps every point of X to a **nonempty** subset of Y.

$F : X \to \mathcal{P}(Y)$ is **upper semicontinuous** (u.s.c.) if for every open set $V \subset Y$ the set $F^{-1}(V) = \{x \in X : F(x) \subset V\}$ is open in X.

$F : X \to \mathcal{P}(X)$ is **Hausdorff upper semicontinuous** (h-u.s.c) if $\forall\, x \in X$ and $\forall\, \varepsilon > 0 \; \exists\, \delta(x,\varepsilon) > 0$ such that $\forall\, z \in B(x,\delta) \subset X$ we have $F(z) \subseteq B(F(x),\varepsilon)$ where $B(A,\varepsilon) = \{y \in X \mid \widetilde{h}(y,A) < \varepsilon\}$. A u.s.c. multifunction is always h-u.s.c. and the two concepts coincide when F has compact values.

We shall say that F is **compact h-u.s.c.** if F is h-u.s.c. and maps bounded sets to precompact sets.

A **selection** of a multifunction $F : X \to \mathcal{P}(Y)$ is a function $g : X \to Y$ such that $g(x) \in F(x) \quad \forall\, x \in X$.

Given a metric space X, the **support** of a function $g : X \to \mathbb{R}$ is the following set:

$$\operatorname{supp} g \equiv \overline{\{x \in X : g(x) \neq 0\}}.$$

A family of continuous functions $\{g_i\}_{i \in I}$ with $g_i : X \to [0,1]$ is a **Lipschitzian partition of unity** if the following conditions are satisfied:

a) the sets $\{\operatorname{supp} g_i\}_{i \in I}$ form a locally finite cover of X, that is every point of X has a neighborhood which intersects a finite number of these supports;

b) $\sum_{i \in I} g_i(x) = 1, \quad \forall\, x \in X$;

c) the functions g_i are Lipschitzian and non-negative.

2. Peano's Theorem in Infinite Dimensional Spaces

Let X be a Banach space and $f(t,x)$ a continuous function from $[t_0, t_0 + a] \times X$ into X. Consider the Cauchy problem:

$$\begin{cases} \dot{x}(t) = f(t, x(t)) \\ x(t_0) = x_0 \end{cases} \qquad \text{C)}$$

where $t \in [t_0, t_0 + a]$ and $x_0 \in X$.

A function $x : [t_0, t_0 + \delta] \to X$ is said to be a solution of C) if:

(i) $x(t_0) = x_0$

(ii) $x(t)$ is **strongly differentiable** in t for $t_0 \leq t \leq t_0 + \delta$ and satisfies $\dot{x}(t) = f(t, x(t))$ on that interval. (Strong differentiability implies that x is continuous in the strong topology, hence $\dot{x}(t)$ is continuous.)

It is known that the problem C) is equivalent to the integral equation:

$$x(t) = x_0 + \int_{t_0}^{t} f(s, x(s)) \, ds$$

where the integral is in the sense of Bochner (see Diestel and Uhl [47]).

The classical theorem which states existence and uniqueness of solutions of C) in the case $X = \mathbb{R}^n$ remains substantially unchanged in infinite dimensional spaces (see [83]).

Theorem 2.1. *Given the rectangle $R_0 = \{(t,x) \in \mathbb{R} \times X : |t - t_0| \leq a, \|x - x_0\| \leq b\}$, let $f : R_0 \to X$ be a continuous function in t for each fixed x. Suppose that $\|f(t, x_1) - f(t, x_2)\| \leq K\|x_1 - x_2\|$ for $(t, x_1), (t, x_2) \in R_0$, where $K \in \mathbb{R}^+$. Let $M = \sup_{R_0} \|f(t, x)\|$.*

Then there exists one and only one solution of the problem C), and its interval of existence is at least $|t - t_0| \leq a'$, where $a' = \min\{a, \frac{b}{M}\}$.

Proof. Obviously the Lipschitz condition plus the continuity in t implies that $f(t, x)$ is continuous in (t, x).

We shall apply the method of successive approximations in the space $C((t_0 - a', t_0 + a'), X)$ in order to prove the existence of a solution of C). Define:

$$\begin{cases} x_0(t) = x_0 \\ x_n(t) = x_0 + \int_{t_0}^{t} f(s, x_{n-1}(s))ds, \text{ with } |t - t_0| \leq a'. \end{cases}$$

By induction the function $x_n(t)$ exists on $|t_0 - t| < a'$ and is strongly continuous. Also, $\|x_n(t) - x_0\| = \left\|\int_{t_0}^{t} f(s, x_{n-1}(s))ds\right\| \leq M|t - t_0| \leq b$ for $|t - t_0| \leq a'$ and

$$\|x_n(t) - x_{n-1}(t)\| \leq \int_{t_0}^{t} K\|x_{n-1}(s) - x_{n-2}(x)\|ds \leq K|t - t_0|\|x_{n-1} - x_{n-2}\|$$

$$\leq K^{n-1}\left(\frac{|t-t_0|^{n-1}}{n!}\right)\|x_1 - x_0\| \leq MK^{n-1}\frac{(a')^n}{n!}.$$

It follows that $\{x_n(t)\}_{n \in N}$ is Cauchy and converges uniformly to a continuous function $x(t)$. Hence, as $n \to \infty$:

$$\|f(t, x_n(t)) - f(t, x(t))\| \leq K\|x_n(t) - x(t)\| \to 0 \quad \text{uniformly in} \quad |t - t_0| \leq a'.$$

By considering the sequence of functions $\{f(\cdot, x_n(\cdot))\}_{n \in N}$ we obtain:

$$x(t) = \lim_{n \to \infty} x_n(t) = x_0 + \lim_{n \to \infty} \int_{t_0}^{t} f(s, x_{n-1}(s))ds = x_0 + \int_{t_0}^{t} f(s, x(s))ds.$$

From the continuity of $f(t, x)$ it follows immediately that $x \in C^1$ in $|t - t_0| \leq a$, so $x(t)$ is a solution of the problem C).

Finally we prove uniqueness. Let $x(t)$ and $y(t)$ be two solutions of C). Then:

$$\|x(t) - y(t)\| = \left\|\int_{t_0}^{t} f(s, x(s)) - f(s, y(s))ds\right\|$$

$$\leq K \int_{t_0}^{t} \|x(s) - y(s)\| ds$$

and by Gronwall's Lemma we have $\|x(t) - y(t)\| \equiv 0$ that is $x = y$.

q.e.d.

Observe that Theorem 2.1 is only a local result, because the classical theorem on continuation of solutions cannot be extended to infinite dimensional spaces. It may happen that a solution of C) exists only in $[t_0, t_0+\delta)$ for some $\delta < a$ and remains bounded there, a situation impossible for $X = \mathbb{R}^n$. The first counterexample is due to Dieudonné [48] in the space c_0. Later on many other examples were obtained, for example, by Lakshmikantham and Ladas [85], Horst [65] and Bourbaki [15] in, respectively, c_0, $L^2[0,1]$ and c_0.

Now let us examine conditions that only ensure the existence of solutions of the problem C). It is well known that, when $X = \mathbb{R}^n$, the continuity of $f(t,x)$ in a neighborhood of (t_0, x_0) implies the existence of local solutions of C). This is the classical Peano theorem which was discussed in the Introduction. This theorem cannot be extended to abstract spaces, as we can deduce from an example given by Dieudonné in 1950 (see [48]).

Example 2.1. Let $X = c_0$ be the space of real sequences $x = \{x_n\}_{n \in N}$ such that $\lim_{n \to \infty} x_n = 0$. Define a function $f : X \to X$ as:

$$f(x) = \{\sqrt{|x_n|} + \tfrac{1}{n}\}_{n \in N}, \quad \forall\, x \in X.$$

The continuity of the scalar function $\sqrt{|x|}$ implies the continuity of $f(x)$, $\forall\, x \in X$. Nevertheless we can show that the problem:

$$\begin{cases} \dot{x}(t) = f(x(t)) \\ x(0) = 0 \end{cases}$$

has no solutions in X. If fact if $x(t) = \{x_n(t)\}_{n \in N}$ were a solution of C), its nth coordinate would satisfy the scalar differential equation:

$$\dot{x}_n(t) = \sqrt{|x_n(t)|} + \tfrac{1}{n} \quad \text{and} \quad x_n(0) = 0.$$

Since the derivative of $x_n(t)$ is strictly positive and $x_n(0) = 0$, we would have $x_n(t) > 0$ for $0 < t < \tau$ where τ is sufficiently small. Hence:

$$\dot{x}_n(t) = \sqrt{x_n(t)} + \frac{1}{n} > \sqrt{x_n(t)} \quad \forall\, t \in (0, \tau).$$

This implies

$$x_n(t) \geq \frac{t^2}{4} \quad \text{for} \quad 0 \leq t < \tau \quad \text{and} \quad \forall\, n \in N.$$

It is clear that the sequence $\{x_n(t)\}_{n \in N}$ cannot converge to zero as $n \to \infty$, that is $x(t) \notin X$, so $x(t)$ cannot be a solution of the initial problem.

At first glance it could seem that the failure of Peano's theorem in the above example depends on the properties of the space c_0, which is not reflexive. However, twenty years later Yorke [158] constructed a counterexample in ℓ_2, which was simplified by Lasota-Yorke [86] to an example in the space $L^2[0, +\infty)$. In 1972 Godunov [54] obtained an independent example in ℓ_2; the following year Cellina [24] succeeded in proving that Peano's theorem is false in every nonreflexive Banach space.

Finally in 1975 Godunov [56] definitively settled the problem with a theorem based on the following result due to Day [32]:

Theorem 2.2. *Let X be an infinite dimensional Banach space. Then there exist two biorthogonal sequences $\{e_n\}_{n \in N}$ and $\{\alpha_n\}_{n \in N}$ in X and X^* respectively, such that:*
  (i) $\|e_n\| = \|\alpha_n\| = 1 \quad \forall\, n \in N$.
  (ii) $\{e_n\}_{n \in N}$ *is a Schauder basis of the closed linear subspace L of X generated by the set $\{e_n\}_{n \in N}$.*
  (iii) *If $P_m(x) = \sum_{n=1}^{m} \alpha_n(x) e_n$, then P_m is a projector from X to L such that $\|P_m\| \leq 1 + \frac{1}{m}$ for $m \in N$.*

Godunov proved the following:

Theorem 2.3. *Peano's theorem is true in a Banach space X if and only if X is finite dimensional.*

Proof. The "if" part of the proof is equivalent to Peano's theorem in $\mathbb{R}^n$. Conversely, assume that Peano's theorem is true in some infinite dimensional space X. Let L, $\{e_n\}_{n \in N}$ and $\{\alpha_n\}_{n \in N}$ be respectively a subspace and two sequences as described in Theorem 2.2. Define:

$$a_n = \frac{1}{2n+1} \qquad b_n = \frac{1}{2n} \qquad c_n = \frac{a_n + b_n}{2}.$$

Let $\{\psi_n(t)\}_{n \in N}$ and $\{\varphi_n(t)\}_{n \in N}$ be two sequences of scalar functions and let $h: \mathbb{R} \to \mathbb{R}$ such that:

$$\psi_n(t) = \begin{cases} 0 & \text{if } t \leq c_n \\ 1 & \text{if } t \geq b_n \\ \text{is linear} & \text{if } c_n < t < b_n \end{cases} \qquad h(u) = \begin{cases} 0 & \text{if } u < 0 \\ u & \text{if } u \in [0,1] \\ 1 & \text{if } u > 1 \end{cases}$$

$\varphi_n(t) = 0$ if $t \notin (a_n, c_n)$, and

$0 < \varphi_n(t) < \dfrac{1}{n}$ with $\varphi_n(t)$ a continuous function, otherwise.

Moreover we introduce three functions P, φ, f from $\mathbb{R} \times L$ into L by means of the following formulas:

$$\varphi(t,x) = \sum_{n=1}^{\infty} \varphi_n(t) h\left(\frac{(t-b_{n+1})^2}{4} - \|x - P_n(x)\|\right) e_n,$$

$$P(t,x) = \sum_{n=1}^{\infty} \psi_n(t) a_n(x) e_n,$$

$$f(t,x) = \begin{cases} \dfrac{P(t,x)}{\sqrt{\|P(t,x)\|}} + \varphi(t) & \text{if } P(t,x) \neq 0, \\ \varphi(t,x) & \text{if } P(t,x) = 0. \end{cases}$$

The functions φ, P, f are continuous. Applying Dugundji's theorem we can extend the function $f : \mathbb{R} \times L \to L$ to all of X preserving the convex hull of the set of values of f. If $g : \mathbb{R} \times X \to L$ denotes this continuous extension, we shall show that the Cauchy problem:

$$\begin{cases} \dot{x}(t) = g(t, x(t)) \\ x(0) = 0 \end{cases}$$

has no solution.

Assume that $x : \mathbb{R} \to X$ is a solution defined in a neighborhood of $t = 0$. Then:

$$x(t) = \int_0^t g(s, x(s)) ds \quad \text{with} \quad g(s, x(s)) \in L \quad \text{so that} \quad x(t) \in L.$$

Since f and g coincide on L, $x(t)$ is also a solution of the initial value problem:

$$\begin{cases} \dot{x}(t) = f(t, x(t)) \\ x(0) = 0. \end{cases}$$

At this point two cases are possible:
a) $\forall n_0 \in N \; \exists n > n_0$ such that $x(b_n) \neq 0$
b) for all sufficiently large n we have $x(b_n) = 0$.
First we consider the case a).

Let $m \in N$ such that $x(b_m) \neq 0$. Obviously $x(t)$ is also a solution of the problem:

$$\begin{cases} \dot{u}(t) = f(t, u(t)) \\ u(b_m) = x(b_m). \end{cases} \qquad \tilde{C})$$

Let us verify that the unique solution of $\tilde{C})$ on $[b_m, 1]$ is the following function:

$$u(t) = \frac{x(b_m)}{\|x(b_m)\|} \cdot \frac{(t - \tilde{b}_m)^2}{4} \quad \text{with} \quad \tilde{b}_m < b_m \quad \text{and} \quad (b_m - \tilde{b}_m)^2 = 4\|x(b_m)\|.$$

It is clear that $u(b_m) = x(b_m)$.

Now let $n < m$, so at most $n = m - 1$ and in this case we obtain:

$$b_m = \frac{1}{2m} < a_n = \frac{1}{2(m-1)+1} = \frac{1}{2m-1} < b_n = \frac{1}{2m-2}.$$

In this case it is clear that b_m is on the left of the interval $[c_n, +\infty)$.

If $n < m$, we have $\psi_n(t) = 0$ for $t < b_m$ and $P(t, x) = \sum_{n=m}^{\infty} \psi_n(t) \alpha_n(t) e_n$. With a similar argument about intervals, we get: $\varphi_n(t) = 0$ for $t < b_m$ when $n < m$, and $\varphi(t, x) = \sum_{n=m}^{\infty} \varphi_n(t) h\left(\frac{(t - b_{n+1})^2}{4} - \|x - P_n(t)\|\right) e_n$. Therefore, if $P(t, x) \neq 0$, $\forall t < b_m$ and $n < m$ we have: $\alpha_n(f(t, x)) = \frac{\alpha_n(P(t,x))}{\sqrt{\|P(t,x)\|}} + \alpha_n(\varphi(t, x)) = 0$ since e_n and α_n are biorthogonal sequences. By similar arguments, if $P(t, x) = 0$ we get:

$$\alpha_n(x(b_m)) = \alpha_n\left(\int_0^{b_m} f(t, x(t))dt\right) = 0 \quad \text{and} \quad \alpha_n(u(t)) = 0 \quad \forall t \in \mathbb{R}$$

by the definition of $u(t)$. As $\psi_n(t) = 1$ on $(b_m, +\infty)$ for $n \geq m$, we have:

$$P(t, u(t)) = \sum_{n=m}^{\infty} \alpha_n(u(t)) e_n = u(t) - \sum_{n=1}^{m-1} \alpha_n(u(t)) e_n = u(t) \quad \text{for} \quad t > b_m.$$

Applying an analogous procedure, we see that $\varphi(t, u(t)) = 0$ for $t \geq b_m$.
Now we have:

$$f(t, u(t)) = \frac{u(t)}{\sqrt{\|u(t)\|}} = \frac{x(b_m)}{\|x(b_m)\|} \cdot \frac{(t - \tilde{b}_m)^2}{4} \cdot \frac{2}{|t - \tilde{b}_m|} = \dot{u}(t) \quad \text{for} \quad t \geq b_m.$$

11

Therefore $u(t)$ is a solution of $\widetilde{C})$.

We now prove the uniqueness of solutions of $\widetilde{C})$.

Let $M = \{(t, u(t)) : t \in [b_m, 1]\} \subset \mathbb{R} \times L$. We want to verify that f is locally Lipschitzian in some neighborhood of the set M.

Given $t \in (0, 1]$, suppose $t \in [b_{n+1}, b_n]$. We shall have the following situation: if $k < n$ then $\psi_k(t) \equiv 0$ on $[b_{n+1}, b_n]$, if $k > n$ $\psi_k(t) \equiv 1$ on $[b_{n+1}, b_n]$ and finally, for $k = n$, $\psi_n(t)$ is a linear function on $[c_n, b_n] \subset [b_{n+1}, b_n]$. Hence:

$$\|P(t,x) - P(t,y)\| = \left\|\sum_{k=1}^{\infty} \psi_k(t)\alpha_k(x-y)e_k\right\|$$

$$\leq \|\psi_n(t)\alpha_n(x-y)e_n\| + \left\|\sum_{k=n+1}^{\infty} \alpha_k(x-y)e_k\right\|$$

$$= \|\psi_n(t)\alpha_n(x-y)e_n\| + \left\|\sum_{k=1}^{\infty} \alpha_k(x-y)e_k\right\|$$

$$\leq \|\psi_n(t)\alpha_n(x-y)e_n\| + \|(x-y) - P_n(x-y)\|$$

$$\leq 4\|x-y\|.$$

As $P(t,x) = u(t) \neq 0$ in some neighborhood of M, it follows that $\frac{P(t,x)}{\sqrt{\|P(t,x)\|}}$ is locally Lipschitzian in the same interval. Also, since $\varphi(t,x) = 0$ in a neighborhood of M, f is locally Lipschitzian in the intersection of these two neighborhoods and $u(t)$ is the unique solution of $\widetilde{C})$. So $x(t) \equiv u(t)$ for $t \geq b_m$.

If $t > b_m$, then $\forall n < m$ we have $\alpha_n(x(t)) = \alpha_n(u(t)) = 0$.

But m can be chosen arbitrarily large, so we have $\alpha_n(x(t)) = 0 \quad \forall n \in N$. Since $\{e_n\}_{n \in N}$ is a Schauder basis, the estimate of $x(t)$ made by means of the functionals α_n implies $x(t) \equiv 0$, contradicting a).

We consider case b).

Let m be such that $x(b_m) = 0$. If $t \in [b_{m+1}, b_m]$, we have:

$$\varphi(t,x) = \varphi_m(t) h\left(\frac{(t-b_{m+1})^2}{4} - \|x - P_m(x)\|\right) e_m \quad \text{and}$$

$$P(t,x) = \sum_{i=m}^{\infty} \psi_i(t)\alpha_i(x)e_i$$

$$\alpha_n(\varphi(t,x)) = 0 \quad \text{and} \quad \alpha_n(P(t,x)) = \alpha_n(x(t)) \quad \forall n > m.$$

12

So $\frac{d}{dt}\alpha_n(x(t)) = \alpha_n(f(t,x(t))) = \frac{\alpha_n(x(t))}{\sqrt{\|P(t,x)\|}}$ $\forall n > m$ and $t \in [b_{m+1}, b_m]$. Now assume that there exists a number $n > m$ such that $\alpha_n(x(t_0)) \neq 0$ for some $t_0 \in [b_{m+1}, b_m]$. From a study of the derivative of $\alpha_n(x(t))$ we have the following cases:

if $\alpha_n(x(t_0)) > 0$ then $0 = \alpha_n(x(b_m)) > \alpha_n(x(t_0)) = 0$.
if $\alpha_n(x(t_0)) < 0$ we have $0 = \alpha_n(x(b_m)) < \alpha_n(x(t_0)) = 0$.

In both cases we obtain $\alpha_n(x(t)) \equiv 0$ on $[b_{m+1}, b_m]$ for $n > m$. If $t \in [0, c_m]$, the following is true:

$$\|x(t) - P_m(x)\| = \left\|\sum_{n=1}^{\infty} \alpha_n(x)e_n - \sum_{n=1}^{m} \alpha_n(x)e_n\right\| = \left\|\sum_{n=m+1}^{\infty} \alpha_n(x)e_n\right\| = 0$$

and also

$$\alpha_m(P(t,x)) = \alpha_m\left(\sum_{n=m+1}^{\infty} \psi_n(t)\alpha_n(x)e_n\right) = 0.$$

Hence

$$\frac{d}{dt}\alpha_n(x(t)) = \varphi_m(t) h\left(\frac{(t-b_{m+1})^2}{4}\right).$$

From the definition of $\varphi_m(t)$ we have $\frac{d}{dt}\alpha_m(x) = 0$ on $[0, a_m]$, so:

$$\frac{d}{dt}\alpha_m(x(t)) = \begin{cases} 0 & \text{on } [0, a_m] \\ > 0 & \text{on } (a_m, c_m). \end{cases}$$

Consequently $\alpha_m(x(c_m)) > 0$ and $\alpha_m(x(b_m)) > 0$, but this contradicts b).

q.e.d.

It is interesting to observe that Godunov [55] has shown that in infinite dimensional spaces Peano's theorem is false even in its "weakened" form, that is, he constructed in the space ℓ_2 a continuous function $f : \mathbb{R} \times \ell_2 \to \ell_2$ such that the problem C) has no solution $\forall (t_0, x_0) \in \mathbb{R} \times \ell_2$. It follows that the failure of Peano's theorem does not depend on the couple (t_0, x_0) of initial data.

Now let us examine the reasons for this dramatic failure. For an interval J and a point $t_0 \in J$, consider the integral operator $F : C(J, X) \to C(J, X)$ defined as:

$$F(x)(t) = x_0 + \int_{t_0}^{t} f(s, x(s))\, ds.$$

It is clear that fixed points of F and solutions of the problem C) coincide exactly. The operator F is obviously continuous and maps bounded sets to bounded sets which are in fact precompact if X is finite dimensional. In this case Schauder's fixed point theorem states that F has at least one fixed point and so problem C) admits a solution. But if X is an abstract space, F is not necessarily compact and we cannot conclude anything about the existence of fixed points.

Therefore the ultimate cause of the failure of Peano's theorem is that *in an infinite dimensional space X, bounded sets need not be precompact in the strong topology of X*, so that operators of the form of F need not have fixed points. Nevertheless there exist particular abstract spaces equipped with particular topologies in which bounded sets are precompact. For example, this is the case for reflexive Banach spaces endowed with the weak topology σ induced by the dual space (see Dunford-Schwartz [51]). So it is an obvious question to ask whether Peano's theorem holds in those spaces, and we shall study this problem for the more general case of locally convex spaces with the minimal assumption of sequential completeness. This hypothesis is necessary, since otherwise we can construct continuous functions that are not integrable; clearly with these functions problem C) cannot admit solutions. The following form of the Banach-Mackey theorem will be useful:

Theorem 2.4. *Let X be a locally convex space, $B \subset X$ a barrel and $K \subset X$ a closed, convex, bounded, balanced and sequentially complete set.*

Then B absorbs K.

Now it is possible to prove the following result due to Kari [74] (1982):

Theorem 2.5. *Let X be a locally convex space, $B \subset X$ a barrel and $(t_0, x_0) \in [a, b) \times X$. Suppose that $f : \mathbb{R} \times X \to X$ is a continuous function on $R = [a, b] \times \{x_0 + B\}$ and $\overline{co}\,\{f(R)\}$ is a compact set.*

Then the problem C) has a solution defined on $[t_0, t_0 + \delta]$ for some $\delta > 0$.

Proof. Let D be the balanced hull of the set $\overline{co}\,\{f(R)\}$. Since the balanced hull of a convex compact set is still convex and compact (see [66]), by Theorem 2.5 B absorbs D and also $\overline{co}\,\{f(R)\} \subset D$, so we can choose a number $\delta > 0$ such that

$[t_0, t_0 + \delta] \subset [a, b]$ and $\delta \cdot \overline{co}\,\{f(R)\} \subset B$. Let $I \equiv [t_0, t_0 + \delta]$ and $\varepsilon \in \mathbb{R}$ with $0 < \varepsilon \leq \delta$. Set:

$$x_\varepsilon(t) = \begin{cases} x_0 & \text{for } t \in [t_0, t_0 + \varepsilon] \\ x_0 + \int_{t_0}^{t-\varepsilon} f(s, x_\varepsilon(s))\,ds & \text{for } t \in [t_0 + \varepsilon, t_0 + \delta]. \end{cases}$$

As f is continuous on R and $\overline{co}\,\{f(R)\}$ is compact (hence sequentially complete), the function f is integrable and $x_\varepsilon : I \to X$ is well defined. Moreover we have:

$$x_0 = \lim_{t \uparrow t_0 + \varepsilon} x_\varepsilon(t) = \lim_{t \downarrow t_0 + \varepsilon} x_\varepsilon(t),$$

that is $x_\varepsilon(t)$ is continuous on I.

Applying the integral mean value theorem we obtain:

$$x_\varepsilon(t) = x_0 \in B \quad \text{for } t \in [t_0, t_0 + \varepsilon]$$

$$x_\varepsilon(t) = x_0 + \int_{t_0}^{t-\varepsilon} f(s, x_\varepsilon(s))\,ds \in x_0 + (t - \varepsilon - t_0)\,\overline{co}\,\{f(R)\}$$

$$\subset x_0 + [(t_0 + \delta) - \varepsilon - t_0]\,\overline{co}\,\{f(R)\}$$

$$= x_0 + (\delta - \varepsilon)\,\overline{co}\,\{f(R)\}$$

$$\subset x_0 + \delta \cdot \overline{co}\,\{f(R)\} \subset x_0 + B \quad \text{for } t \in [t_0 + \varepsilon, t_0 + \delta].$$

So $x_\varepsilon(t) \subset x_0 + B\ \forall t \in I$. Similarly we have:

$$x_\varepsilon(t) - x_\varepsilon(s) \in (t - s)\,\overline{co}\,\{f(R)\} \quad \forall\, t, s \in I.$$

Now consider a decreasing sequence $\{\varepsilon_n\}_{n \in N}$ with $\varepsilon_1 < \delta$ such that $\lim_{n \to \infty} \varepsilon_n = 0$.

Since $|x_\varepsilon(t) - x_\varepsilon(s)| \to 0$ as $|t - s| \to 0$ with $t, s \in I$, uniformly in ε, the family of continuous functions $H = \{x_{\varepsilon_n}\}_{n \in N}$ will be equicontinuous and the set $H(t) = \{x_{\varepsilon_n}(t)\}_{n \in N}$ will be precompact in $X\ \forall\, t \in I$. Under these conditions the Ascoli-Arzelà theorem implies that H is precompact in the space $C(I, X)$. Therefore H has an accumulation point $x \in C(I, X)$. We want to show that $x(t)$ is a solution of problem C) on I.

Let $U \subset X$ be some barrelled neighborhood of zero. The set $G = I \times \{x_0 + \delta D\}$ is compact, so that the restriction of f to G is uniformly continuous. Then there exists a barrelled neighborhood V of zero where the following relation is true:

$$\text{if } (s,x) \text{ and } (s,y) \in G \text{ with } x - y \in V, \text{ then } f(s,x) - f(s,y) \in U.$$

Since x is an accumulation point of H and $\overline{co}\ \{f(R)\}$ is compact, we can find a number $n \in N$ such that:

a) $\varepsilon_n \cdot \overline{co}\ \{f(R)\} \subset U$ b) $x(s) - x_{\varepsilon_n}(s) \in V \cap U$ for $\forall s \in I$.

Hence:

$$x(t) - x_0 - \int_{t_0}^{t} f(s, x(s))\, ds$$

$$= x(t) - \int_{t_0}^{t} f(s, x(s))\, ds - x_{\varepsilon_n}(t) + \int_{t_0}^{t-\varepsilon_n} f(s, x_{\varepsilon_n}(s))\, ds$$

$$= [x(t) - x_{\varepsilon_n}(t)] + \int_{t_0}^{t} [f(s, x_{\varepsilon_n}(s)) - f(s, x(s))]\, ds$$

$$- \int_{t-\varepsilon_n}^{t} f(s, x_{\varepsilon_n}(s))\, ds \in U \cap V + (t - t_0)U$$

$$- \varepsilon_n \cdot \overline{co}\ \{f(R)\} \subset U + \delta U - U \subset (2 + \delta)U \quad \forall t \in I.$$

But U is arbitrary, so we have:

$$x(t) = x_0 - \int_{t_0}^{t} f(s, x(s))\, ds \quad \text{for} \quad t \in I.$$

q.e.d.

Now assume that the locally convex space X is sequentially complete and X contains a compact barrel B_0. In this case closed and bounded sets of X are sequentially complete and they can be made closed and balanced by means of respective hulls. By the Banach-Mackey theorem, B_0 absorbs bounded and closed sets which become compact. Consequently, the set $\{f(R)\}$ of the previous theorem is compact and co $\{f(R)\}$ is precompact in X (see Horvath [66]). As the conditions of Theorem 2.5 are satisfied, the assumption of continuity of f and the existence of a compact barrel in X are sufficient to ensure that problem C) has a solution.

The next theorem tells us precisely when the second assumption is true.

Theorem 2.6. *Let E be a linear space. Then E is a locally convex sequentially complete space containing a compact barrel if and only if $E = (X^*, \tau)$ where X^* is the dual of a normed barrelled space and τ is a locally convex topology on X^* such that $\sigma(X^*, X) \subseteq \tau \subseteq \lambda(X^*, X)$.*

Proof. "If": Let X^* be the dual of a normed barrelled space X and τ a topology satisfying the hypotheses. Consider the closed unit ball $B_1 = \{x^* \in X^* : \|x^*\| \leq 1\}$ which is clearly barrelled in X^*.

Applying the Banach-Alaoglu Theorem, we see that B_1 is compact in the weak topology $\sigma(X^*, X)$ and, by the Banach-Dieudonné Theorem, the topology $\lambda(X^*, X)$ of precompact convergence is the strongest topology which makes B_1 compact. Consequently, B_1 will be compact also in $(X^*, \tau) = E$.

It remains to show the sequential completeness of E.

Let $K \subset X^*$ be a bounded set in the τ-topology. Since τ is finer than $\sigma(X^*, X)$, K is σ-bounded and so equicontinuous (see [66]). Hence K is bounded in the norm topology $\beta(X^*, X)$ and there exists a constant $r \in \mathbb{R}$ such that $K \subset rB_1$.

But rB_1 is compact, so K is τ-totally bounded. From the arbitrary choice of K we see that E is sequentially complete.

Conversely, let E be a locally convex space containing a compact barrel B. We denote by τ_0 the original topology of E. Applying Theorem 2.4 to the convex and balanced hull of E, it follows that B absorbs τ_0-bounded sets in E. So the closed bounded sets are compact in (E, τ_0) and E becomes semireflexive (see [66]).

For the space X we choose $X = (E^*, \beta(E^*, E))$ where $\beta(E^*, E)$ represents the strong topology on E^* induced by E.

Let $A \subset X$ be a bounded set and $U_{\varepsilon, A} = \{x^* \in E^* : |\langle x, x^* \rangle| \leq \varepsilon \; \forall x \in A\}$ an arbitrary strong neighborhood of zero. Since B absorbs bounded sets, there exists $\delta \in \mathbb{R}$ such that $\delta A \subset B$. Then $\frac{a}{\varepsilon \delta} U_{\varepsilon, A}$ is still a strong neighborhood of X and we have:

$$\frac{1}{\varepsilon} U_{\varepsilon, A} = \{\frac{1}{\varepsilon} x^* \in E^* : |\langle x, x^* \rangle| \leq \varepsilon \;\; \forall x \in A\}$$

$$= \{y^* \in E^* : |\langle x, \varepsilon y^* \rangle| \leq \varepsilon \;\; \forall x \in A\}$$

$$= \{y^* \in E^* : |\langle x, y^* \rangle| \leq 1 \;\; \forall x \in A\} \quad \text{and}$$

$$\delta B = (\frac{1}{\delta} B)^0 = \{x^* \in E^* : |\langle x, x^* \rangle| \leq 1 \;\; \forall x \in \frac{1}{\delta} B\}$$

where $B^0 \subset E^*$ stands for the polar set of $B \subset E$.

As $A \subset \frac{1}{\delta} B$, we have $\delta B^0 \subset \frac{1}{3} U_{\varepsilon,A}$ and $B^0 \subset \frac{1}{\varepsilon\delta} U_{\varepsilon,A}$. Therefore B^0 is a bounded neighborhood of zero in the Hausdorff space X; that is, X is normable (see [66]). It follows that $X^* = (E^{**}, \beta(E^{**}, E^*))$ is a Banach space and the semireflexivity of E implies:

$$(E, \tau_0) = (X^*, \tau).$$

Now we must prove that the τ-topology is included between $\sigma(X^*, X)$ and $\lambda(X^*, X)$.

First we observe that the space X is the dual of (X^*, τ), so the weak topology $\sigma(X^*, X)$ induced by X is weaker than the original τ-topology.

By the Banach-Dieudonné theorem we know that $\lambda(X^*, X)$ is the strongest of those topologies equivalent to $\sigma(X^*, X)$ on the bipolar $B^{00} \subset X^*$ and we have $B^{00} \equiv B$ by the theorem on bipolars (see [66]). Considering the compactness of B, it follows that the restriction to B of the identity function $Id : (X^*, \tau) \to (X^*, \sigma(X^*, X))$ is a homeomorphism. So τ induces on B a topology equivalent to $\sigma(X^*, X)$. Hence:

$$\sigma(X^*, X) \subseteq \tau \subseteq \lambda(X^*, X).$$

It remains to show that X is a barrelled space.

Since X is the dual of (X^*, τ), every $\sigma(X^*, X)$-bounded set in X^* is τ-bounded. Using the assumption of sequential completeness of (X^*, τ) together with Theorem 2.4, we see that B absorbs τ-bounded sets which therefore are precompact and bounded in norm. Thus $\sigma(X^*, X)$-bounded sets of X are also $\beta(X^*, X)$-bounded and, as the opposite is always true, strong boundedness is equivalent to weak boundedness on X^*. Consequently, X is barrelled (see [66]).

q.e.d.

From Theorems 2.5 and 2.6 follows immediately:

Theorem 2.7. *Let X be a normed barrelled space, τ a locally convex topology on X^* satisfying $\sigma(X^*, X) \subseteq \tau \subseteq \lambda(X^*, X)$ and $E = (X^*, \tau)$. Assume that $f : \mathbb{R} \times E \to E$ is continuous and $(t_0, x_0) \in \mathbb{R} \times E$.*

Then the problem C) admits a local solution.

This theorem contains an important earlier result due to Szép [130] (1971). He showed that Peano's theorem holds in every reflexive Banach space endowed with the weak topology. Under these assumptions X becomes the dual of X^* which, since it is normed and complete, is barrelled; therefore the hypotheses of Theorem 2.6 are valid and problem C) has a solution.

Later Kari [74], generalizing a construction due to Cellina (see [24]), proved that in every nonsemireflexive quasi-complete locally convex space we can find a continuous function such that problem C) does not admit any solution. In other words semireflexivity is a necessary condition for the validity of Peano's theorem.

From this fact it is possible to prove the converse of Szép's theorem, i.e.,

Theorem 2.8. *Let X be a Banach space. Then X is reflexive if and only if Peano's theorem holds in the space X equipped with the weak topology $\sigma(X, X^*)$.*

This result represents an unusual characterization of reflexivity in a Banach space.

In conclusion, we note the results of Lasota and Yorke [86] and Vidossich ([153], [154]) on the generic nature of existence and continuous dependence for initial-value problems in a Banach space. Lasota and Yorke showed that existence, uniqueness and continuous dependence are generic, in the sense that the set

{f | there exists a unique solution to the initial value problem, it is "unlimited", and solutions depend continuously at (t_0, x_0) on f and the initial conditions}

has complement of category 1. The topology for the space containing f is that of uniform convergence on bounded sets. Here "unlimited" means that if the maximal interval of existence is (a, b), then $\lim_{t_n \downarrow a} x(t_n)$ and $\lim_{t_n \uparrow b} x(t_a)$ do not exist for any sequences $\{t_n\}$. Vidossich presented a shorter proof, based on a sophisticated fixed point result.

3. Kneser Type Theorems

In the seventies, there was considerable interest in the structure of the solution set of a differential equation in an abstract space. This research faced various additional difficulties compared to the finite dimensional case. In the previous chapter we have seen that continuity of f is not sufficient to ensure the existence of solutions of the problem C); therefore the classical Kneser and Hukuhara Theorems concerning the topological structure of "The Peano funnel" in $\mathbb{R}^n$ cannot be extended to infinite dimensional spaces.

Moreover, even the assumption of existence of solutions does not allow us to conclude anything about properties of the solution set; witness the following example taken from Cellina [22] (1971):

Example 3.1. Let $X = \ell^\infty$ be the Banach space of real bounded sequences $x = \{x_n\}_{n \in \mathbb{N}}$ with norm $\|x\|_\infty = \sup_{n \in \mathbb{N}} |x_n|$. Define the function $f : X \to X$ as:

$$f(x) = \begin{cases} 2 \cdot \dfrac{x}{\sqrt{\|x\|_\infty}} & \text{if } x \neq 0 \\ 0 & \text{if } x = 0. \end{cases}$$

As $\lim_{x \to 0} \|f(x)\|_\infty = 0$, f will be continuous on all of X.

Consider the following Cauchy problem:

$$\begin{cases} \dot{x}(t) = f(x) \\ x(0) = 0 \end{cases} \qquad \text{C*)}$$

If we set this differential equation in $\mathbb{R}$, its Peano funnel consists of the family of functions of the form:

$$x_\delta(t) = \begin{cases} 0 & \text{if } \alpha \leq t \leq \beta; \\ (t - \beta)^2 & \text{if } t > \beta; \\ (t - \alpha)^2 & \text{if } t < \alpha; \end{cases}$$

where $\alpha \leq 0 \leq \beta$. This family is a continuum in the space $C([0, a], X)$ with $a > 0$.

Similarly, it is simple to verify that all functions of the form:

$$x_1(t) = \begin{cases} 0 & \text{for } \alpha \leq t \leq \beta; \\ ((t-\beta)^2, 0, \ldots, 0, \ldots) & \text{for } t > \beta; \\ ((t-\alpha)^2, 0, \ldots, 0, \ldots) & \text{for } t < \alpha; \end{cases}$$

$$x_n(t) = \begin{cases} 0 & \text{for } \alpha \leq t \leq \beta; \\ (0, \ldots, (t-\beta)^2, 0, \ldots) & \text{for } t > \beta; \\ (0, \ldots, (t-\alpha)^2, 0, \ldots) & \text{for } t < \alpha; \end{cases}$$

are solutions of C*) in X for any $\alpha \leq 0 \leq \beta$ and $n \in N$, where $x_n(t)$ has zero entry except for the nth coordinate.

Other possible solutions must be linear combinations of those above, since otherwise their scalar components could not satisfy the corresponding scalar differential equations.

Let $\{a_n\}_{n \in N}$ be a bounded sequence with real values, fix $\beta > 0$ and define

$$x(t) = a_1 x_1(t) + a_2 x_2(t) + \cdots = \left(a_1(t-\beta)^2, a_2(t-\beta)^2, \ldots\right)$$

be the linear combination of the above solutions for $t > \beta$. Then we have:

$$f(x) = 2 \cdot \frac{(a_1(t-\beta)^2, a_2(t-\beta)^2, \ldots)}{\sqrt{\sup_{n \in N} \{|a_n(t-\beta)^2|\}}} = 2 \cdot \frac{(a_1(t-\beta), a_2(t-\beta), \ldots)}{\sqrt{\sup_{n \in N} |a_n|}}$$

and $\dot{x}(t) = 2 \cdot (a_1(t-\beta), a_2(t-\beta), \ldots)$.

Clearly $\dot{x}(t) = f(x)$ is valid only if:

$$\sqrt{\sup_{n \in N} |a_n|} = 1, \quad \text{that is} \quad \sup_{n \in N} |a_n| = 1.$$

So, in the case $X = \ell^\infty$, Peano's funnel is composed of the family of functions of the form:

$$x(t) = \begin{cases} 0 & \text{for } \alpha \leq t \leq \beta; \\ \left(a_1(t-\beta)^2, a_2(t-\beta)^2, \ldots\right) & \text{for } t > \beta \\ \left(a_1(t-\alpha)^2, a_2(t-\alpha)^2, \ldots\right) & \text{for } t < \alpha \end{cases}$$

where $\alpha \leq 0 \leq \beta$ and $\{a_n\}_{n\in N} \in X$ with $\|a_n\|_\infty = 1$.

Choosing as the sequences $\{a_n\}$ all possible sequences $\{e_n\}_{n\in N}$ with a single component different from zero and setting $\alpha = \beta = 0$ we obtain the subset S_0 of S consisting of the following solutions of C*):

$$x_1(t) = (t^2, 0, 0, \dots), \dots, x_n(t) = (0, \dots, t^2, 0, \dots) \quad \text{where} \quad n \in N.$$

Taking $a > 0$, we can prove that S_0 is not compact in any space $C([0,a], X)$ endowed with the norm $\|x\| = \sup_{t\in[0,a]} \|x(t)\|_\infty$. Let d be the distance in $C([0,a], X)$. Then for $i, j \in N$ with $i \neq j$ we have:

$$d(x_i, x_j) = \sup_{t\in[0,a]} \|x_i(t) - x_j(t)\|_\infty$$

$$= \sup_{t\in[0,a]} \|(0, \dots, t^2, \dots, t^2, 0, \dots)\|_\infty = a^2 \neq 0.$$

Consequently, for $\varepsilon < a^2$ there does not exist a finite ε-net for S_0, i.e., S_0 is not totally bounded. Clearly also the whole solution set S cannot be compact.

Although in this case S is connected, we note that the property of connectedness can also fail in infinite dimensional spaces. We refer the reader to a counterexample constructed by Binding [13] in the space c_0.

From the preceding examples, it is clear that we must add a new condition to the continuity of the function f in order to ensure the existence of solutions of the problem C) and make its Peano funnel topologically "regular." This will be the purpose of the present chapter and we shall start with the case of a Banach space X equipped with the strong topology.

Since the loss of compactness of closed and bounded sets is the main reason for the failure of Peano's theorem, it is natural to "measure" this loss with the Kuratowski or Hausdorff measure of noncompactness. The main results obtained using this approach are due, in chronological order since 1967, to Ambrosetti [2], Sadovski [122], Szufla [131] or [132], Goebel and Rzymowski [57], Kamenskii [71], Cellina [22] or [23] and Chow and Schuur [26]. However they limit themselves to proving existence of solutions of the problem C). One of the first mathematicians who was interested in the structure of the solution set was Deimling in 1975 (see [40]) with an article which was simplified two years later (see [41]). In Deimling's monograph [41] the following theorems were frequently used:

Theorem 3.1. *Let X and Y be two Banach spaces, $A \subset X$ an open set and $f : A \to Y$ a continuous function. Then $\forall\, \varepsilon > 0$ there exists a locally Lipschitzian function $f_\varepsilon : A \to Y$ such that $\sup_{x \in A} \|f(x) - f_\varepsilon(x)\| \leq \varepsilon$.*

Theorem 3.2. *Let $x : [0, a] \to X$ be a differentiable function. Then we have:*

$$\left\{ \frac{x(t) - x(t-h)}{h} : t \in [0, a] \text{ and } 0 < h < t \right\} \subset \overline{co}\,\{\dot{x}(t) : t \in [0, a]\}.$$

These theorems make it possible to prove the existence of approximate solutions for the problem C):

Theorem 3.3. *Let X be a Banach space $I \equiv [0, a]$, $D = B(x_0, r)$ be a closed ball in X and $f : I \times D \to X$ a continuous function with $\|f(t, x)\| \leq M\ \forall\, (t, x) \in I \times D$. Let $\varepsilon > 0$ and $a_\varepsilon = \min\{a, \frac{r}{M+\varepsilon}\}$. Then there exists a differentiable function $x_\varepsilon : [0, a_\varepsilon] \to D$ such that*

$$\dot{x}_\varepsilon(t) = f(t, x_\varepsilon(t)) + y_\varepsilon(t) \quad \text{with} \quad x_\varepsilon(0) = x_0 \quad \text{and} \quad \|y_\varepsilon(t)\| \leq \varepsilon \quad \text{on} \quad [0, a_\varepsilon].$$

Proof. Applying Dugundji's theorem, the function f admits a continuous extension $\widetilde{f} : \mathbb{R} \times X \to X$ such that $\|\widetilde{f}(t, x)\| \leq M$ everywhere. By Theorem 3.1 there exists a locally Lipschitz function $\widetilde{f}_\varepsilon : \mathbb{R} \times X \to X$ such that $\|\widetilde{f}_\varepsilon(t, x) - \widetilde{f}(t, x)\| \leq \varepsilon$.

In particular we have:

$$\|\widetilde{f}_\varepsilon(t, x) - f(t, x)\| \leq \varepsilon \quad \text{and} \quad \|\widetilde{f}_\varepsilon(t, x)\| \leq M + \varepsilon \quad \text{on} \quad I \times D.$$

From Theorem 2.1 it follows that the problem:

$$\begin{cases} \dot{x}(t) = \widetilde{f}_\varepsilon(t, x(t)) \\ x(0) = x_0 \end{cases}$$

has at least a local solution $x_\varepsilon(t)$ defined on $[0, a_\varepsilon]$. It will satisfy the equality:

$$\dot{x}_\varepsilon(t) = f(t, x_\varepsilon(t)) + \left(\widetilde{f}_\varepsilon(t, x_\varepsilon(t)) - f(t, x_\varepsilon(t))\right) = f(t, x_\varepsilon(t)) + y_\varepsilon(t) \quad \text{where}$$

$$\|y_\varepsilon(t)\| = \|\widetilde{f}_\varepsilon(t, x_\varepsilon(t)) - f(t, x_\varepsilon(t))\| \leq \varepsilon.$$

q.e.d.

The next theorem illustrates the main topological properties of the Peano funnel:

Theorem 3.4. *Let X be a Banach space, $D = \overline{B}(x_0, r)$ a closed ball in X, $I \equiv [0, a] \subset \mathbb{R}$, $f : I \times D \to X$ a continuous function with $\|f(t,x)\| \leq M$ on $I \times D$ and $b < \min\{a, \frac{r}{M}\}$ a constant. Suppose that f satisfies the inequality:*

$$\alpha(f(I \times A)) \leq g(\alpha(A)) \quad \forall A \subset D, \qquad \text{a)}$$

where $g : \mathbb{R}^+ \to \mathbb{R}^+$ is continuous and the Cauchy problem $\dot{y}(t) = g(y)$, $y(0) = 0$ has only the trivial solution $y(t) \equiv 0$ on I (in other words, g is a Kamke function).

Then the set S of solutions defined on $[0, b]$ is a continuum in $C(I, X)$. In particular the set $S(t) = \{x(t) : x \in S\}$ is a continuum in X $\forall t \in [0, b]$.

Proof. First we must show that S is nonempty, that is to say the problem C) admits a solution on $[0, b]$. Applying the preceding theorem, let x_n, $n \in N$, be approximate solutions satisfying the equation:

$$\dot{x}_n(t) = f(t, x_n(t)) + y_n(t) \text{text} \quad \text{with} \quad x_n(0) = 0 \quad \text{and} \quad \|y_n(t)\| \leq \frac{1}{n}. \quad (*)$$

We want to prove that the sequence of functions $\{x_n\}_{n \in N}$ is compact in $C([0, b], X)$. Choose a number $\varepsilon > 0$ and set $\delta_\varepsilon = \frac{\varepsilon}{M+1}$. Then for $t_1, t_2 \in [0, b]$ with $|t_1 - t_2| < \delta_\varepsilon$ we have:

$$\|x_n(t_1) - x_n(t_2)\| = \left\| \int_{t_1}^{t_2} [f(s, x_n(s)) + y_n(s)] \, ds \right\|$$

$$\leq \left(M + \frac{1}{n}\right) \cdot |t_1 - t_2| < \varepsilon \quad \forall n \in N.$$

Therefore the sequence $\{x_n\}_{n \in N}$ is equicontinuous. By the Ascoli-Arzelà theorem, we need only prove that the set $\{x_n(t)\}_{n \in N} \subset X$ is totally bounded $\forall t \in [0, b]$, that is $\alpha(\{x_n(t)\}_{n \in N}) = 0$. Set:

$$S_k(t) = \{x_n(t) : n \geq k\} \quad \text{and} \quad \dot{S}_k(t) = \{\dot{x}_n(t) : n \geq k\}.$$

Let $\phi_k : [0, b] \to \mathbb{R}^+$ be the function $\phi_k(t) = \alpha(S_k(t))$. Then

$$\phi_k(0) = \alpha(S_k(0)) = \alpha(\{x_0\}) = 0,$$

and $\forall\, t_1, t_2 \in [0, b]$ we have

$$\phi_k(t_1) - \phi_k(t_2) = \alpha\big(S_k(t_1)\big) - \alpha\big(S_k(t_2)\big).$$

Now we note that

$$\{x_n(t_1): n \geq k\} \subset \{x_n(t_1) - x_n(t_2): n \geq k\} + \{x_n(t_2): n \geq k\}. \qquad (*)$$

Therefore

$$S_k(t_1) \subset \{x_n(t_1) - x_n(t_2) : n \geq k\} + S_k(t_2),$$

and applying α to both sides we obtain

$$\alpha\big(S_k(t_1)\big) - \alpha\big((S_k(t_2)\big) \leq \alpha\left(\{x_n(t_1) - x_n(t_2) : n \geq k\}\right),$$

thus

$$|\varphi_k(t_1) - \varphi_k(t_2)| \leq \alpha\left(\{x_n(t_1) - x_n(t_2) : n \geq k\}\right) \qquad (**)$$

$$\leq 2 \sup_{n \geq k} \|x_n(t_1) - x_n(t_2)\|$$

$$\leq 2|t_1 - t_2|\left(M + \frac{1}{k}\right) \leq 2|t_1 - t_2|(M+1),$$

so ϕ_k is continuous on $[0, b]$. For any fixed t consider:

$$D^-\phi_k(t) = \lim_{\tau \to 0^+} \sup_{\delta \in (0, \tau]} q(\delta) \quad \text{with} \quad q(\delta) = \frac{\phi_k(t) - \phi_k(t-\delta)}{\delta}.$$

Now repeat the argument above, beginning with $(*)$, with $t_1 = t$, $t_2 = t - \delta$, to arrive at:

$$\phi_k(t) - \phi_k(t-\delta) \leq \alpha\big(\{x_n(t) - x_n(t-\delta) : n \geq k\}\big). \qquad (**)'$$

Then $q(\delta) \leq \alpha\big(\{[x_n(t) - x_n(t-\delta)]/\delta \, : n \geq k\}\big)$ and

$$\sup_{\delta \in (0, \tau]} q(\delta) \leq \alpha\left(\left\{\frac{x_n(t) - x_n(t-\delta)}{\delta} : n \geq k,\ 0 < \delta \leq \tau\right\}\right).$$

If we take the limit as $\tau \to 0^+$ and use Theorem 3.2, we get:

$$D^-\phi_k(t) \leq \lim_{\tau \to 0^+} \alpha\big(\overline{\text{co}}\,\{\dot{x}_n(s) : n \geq k,\, s \in I_\tau \equiv [t-\tau, t]\}\big)$$
$$= \lim_{\tau \to 0^+} \alpha\big(\{\dot{x}_n(s) : n \geq k,\, s \in I_\tau\}\big) = \lim_{\tau \to 0^+} \alpha\big(\bigcup_{s \in I_\tau} \dot{S}_k(s)\big).$$

But

$$\alpha\big(\bigcup_{s \in I_\tau} \dot{S}_k(s)\big) = \alpha\big(\bigcup_{s \in I_\tau} \{\dot{x}_n(s) : n \geq k\}\big)$$

$$= \alpha\big(\bigcup_{s \in I_\tau} \{f(s, x_n(s)) + y_n(s) : n \geq k\}\big)$$

$$\leq \alpha\big(\bigcup_{s \in I_\tau} \{f(s, x_n(s)) : n \geq k\}\big) + \alpha\big(\bigcup_{s \in I_\tau} \bigcup_{n \geq k} \{y_n(s)\}\big)$$

$$\leq \alpha\big(\bigcup_{s \in I_\tau} f(s, S_k(s))\big) + \tfrac{2}{k}.$$

By Hypothesis a) we have:

$$\alpha\big(\bigcup_{s \in I_\tau} f(s, S_k(s))\big) \leq \alpha\big(f([0,b] \times \bigcup_{s \in I_\tau} S_k(s))\big) \leq g\big(\alpha\big(\bigcup_{s \in I_\tau} S_k(s)\big)\big).$$

Hence $D^-\phi_k(t) \leq \lim_{\tau \to 0^+} \alpha\big(\bigcup_{s \in I_\tau} f(s, S_k(s))\big) + \tfrac{2}{k} \leq \lim_{\tau \to 0^+} g\big(\alpha\big(\bigcup_{s \in I_\tau} S_k(s)\big)\big) + \tfrac{2}{k}$ and, from the continuity of g and α, $\forall t \in (0, b]$ we have:

$$D^-\phi_k(t) \leq g\big(\alpha\big(\lim_{\tau \to 0^+} \bigcup_{s \in I_\tau} S_k(s)\big)\big) + \tfrac{2}{k} \leq g(\alpha(S_k(t))) + \tfrac{2}{k} = g(\phi_k(t)) + \tfrac{2}{k}. \quad \text{b)}$$

(Note that the Hausdorff distance $h\big(\bigcup_{s \in I_\tau} S_k(s), S_k(t)\big) \leq M\tau$.) Now consider the differential equation:

$$\begin{cases} \dot{y}_n = g(y_n) + \dfrac{1}{n} \\ y_n(0) = \dfrac{1}{n}. \end{cases}$$

Since g is a Kamke function, the sequence of solutions $\{y_n\}_{n \in N}$ converges to zero as $n \to \infty$. Moreover the relation b) implies that $\phi_k(t) < y_n(t)$ for $n < \tfrac{k}{2}$.

So $\lim_{k\to\infty} \phi_k(t) = 0$, but $\forall\, k \in N$ we have:

$$\phi_k(t) = \alpha(\{x_n(t) : n \geq k\}) = \alpha(\{x_n(t) : n \geq 1\}),$$

therefore $\quad \alpha(\{x_n(t) : n \geq 1\}) = \lim_{k\to\infty} \alpha(\{x_n(t) : n \geq k\}) = 0$

Therefore the sequence $\{x_n(t)\}_{n\in N}$ is totally bounded $\forall\, t \in [0,b]$ and $\{x_n\}_{n\in N}$ is precompact in $C([0,b], X)$. Let $\{x_{n_k}\}_{k\in N}$ be a subsequence converging to $\widetilde{x} \in C([0,b], X)$. By the definition of x_n we obtain:

$$\dot{x}_{n_k}(t) = f(t, x_{n_k}(t)) + y_{n_k}(t)$$

and, by passing to the equivalent integral equation and taking the limit as $k \to \infty$, we have:

$$\dot{\widetilde{x}}(t) = f(t, \widetilde{x}(t)).$$

Thus $\widetilde{x}$ is a solution of the problem C), and so S is nonempty.

In order to prove the precompactness of S, it is sufficient to repeat the existence proof, considering a sequence of solutions instead of the sequence $\{x_n\}_{n\in N}$ of approximate solutions. In this way we get a uniformly converging subsequence of solutions whose limit is again a solution. The integral operator:

$$F(x) = x_0 + \int_0^t f(s, x(s))\, ds$$

is continuous on D, so the operator $(Id - F)$ will also be continuous on D. Since for a continuous function the pre-image of a closed set is a closed set, and S is the pre-image of $\{0\}$ for the function $(Id - F)$, S is closed. Consequently, S is compact.

Finally we prove the connectedness of S. Suppose that S is not connected, that is:

$$S = S_1 \cup S_2 \quad \text{with} \quad S_1 \cap S_2 = \emptyset \quad \text{and} \quad S_1,\, S_2 \quad \text{nonempty compact sets.}$$

Let $h(A, B)$ be the Hausdorff distance function for sets; it is clear that $\rho_0 = h(S_1, S_2) > 0$. Define the function $\psi : C([0,b], X) \to \mathbb{R}$ as:

$$\psi(x) = \widetilde{h}(x, S_1) - \widetilde{h}(x, S_2)$$

where $\tilde{h}$ denotes the distance of a point from a set. The function ψ is continuous and we have:

$$\psi(x) \leq -\rho_0 \quad \text{for} \quad x \in S_1 \quad \text{and} \quad \psi(x) \geq \rho_0 \quad \text{for} \quad x \in S_2.$$

Our object is to construct a function $x \in S$ such that $\psi(x) = 0$, a contradiction.

Let $\varepsilon > 0$ be a real number such that $\frac{M+2\varepsilon}{b} < r$. By Theorem 3.1 we can find a function $g_\varepsilon : I \times D \to X$ such that $\|g_\varepsilon(t,x) - g_\varepsilon(t,y)\| \leq k_\varepsilon(t)\|x - y\|$ and $\|g_\varepsilon(t,x) - f(t,x)\| \leq \varepsilon$ on $I \times D$ with $k_\varepsilon(t)$ locally integrable on I. Let $x_1 \in S_1$ and $x_2 \in S_2$ be fixed. Consider three functions f_1, f_2, f^λ from $I \times D$ into X defined as:

$$f_i(t,x) = g_\varepsilon(t,x) + f(t, x_i(t)) - g_\varepsilon(t, x_i(t)) \quad \text{for} \quad i = 1, 2$$

$$f^\lambda(t,x) = f_1(t,x) + \lambda(f_2(t,x) - f_1(t,x)) \quad \text{with} \quad \lambda \in [0,1].$$

If $x, y \in I \times D$, we have:

$$\|f_i(t,x) - f_i(t,y)\| = \|g_\varepsilon(t,x) - g_\varepsilon(t,y)\| \leq k_\varepsilon(t)\|x - y\| \quad \text{for} \quad i = 1, 2.$$

Consequently, the functions f_i are locally Lipschitzian as is the linear combination f^λ. Moreover we have:

$$\|f_i(t,x) - f(t,x)\| = \|g_\varepsilon(t,x) + f(t,x_i) - g_\varepsilon(t,x_i) - f(t,x)\|$$

$$\leq \|g_\varepsilon(t,x) - f(t,x)\| + \|g_\varepsilon(t,x_i) - f(t,x_i)\| \leq 2\varepsilon$$

and

$$\|f^\lambda(t,x) - f(t,x)\| = \lambda\|f(t,x_2) - g_\varepsilon(t,x_2) - f(t,x_1) + g_\varepsilon(t,x_1)\|$$

$$+ \|f_1(t,x) - f(t,x)\|$$

$$\leq 2\varepsilon + \lambda(\varepsilon + \varepsilon) = (2 + 2\lambda)\varepsilon \leq 4\varepsilon.$$

From Theorem 2.1 we deduce that the Cauchy problem:

$$\begin{cases} \dot{x}(t) = f^\lambda(t, x(t)) \\ x(0) = 0 \end{cases} \qquad \text{P)}$$

has a unique solution x^λ in $[0, b] \; \forall \lambda \in [0, 1]$.

If $\lambda_1, \lambda_2 \in [0,1]$, then:

$$\|x^{\lambda_2}(t) - x^{\lambda_1}(t)\|$$

$$= \left\| \int_0^t f^{\lambda_2}(s, x^{\lambda_2}(s)) - f^{\lambda_1}(s, x^{\lambda_1}(s)) ds \right\|$$

$$= |\lambda_2 - \lambda_1| \cdot \left\| \int_0^t f(s, x_2(s)) - g_\varepsilon(s, x_2(s)) - f(s, x_1(s)) + g_\varepsilon(s, x_1(s)) ds \right\| +$$

$$+ \int_0^t |g_\varepsilon(s, x^{\lambda_2}(s)) - g_\varepsilon(s, x^{\lambda_1}(s))| ds$$

$$\leq |\lambda_2 - \lambda_1| b \, 2\varepsilon + \int_0^t k_\varepsilon(s) \|x^{\lambda_2}(s) - x^{\lambda_1}(s)\| ds$$

and by Gronwall's inequality

$$\|x^{\lambda_2} - x^{\lambda_1}\| = \sup_{t \in [0,b]} \|x^{\lambda_2}(t) - x^{\lambda_1}(t)\| \leq 2b|\lambda_2 - \lambda_1|\varepsilon(e^{\int_0^t k_\varepsilon(u)du} - 1).$$

Therefore the function $\varphi : [0,1] \to C([0,b], X)$ such that $\varphi(\lambda) = x^\lambda$ is continuous in $[0,1]$. Also the function $\eta : [0,1] \to \mathbb{R}$ defined as:

$$\eta(\lambda) = \psi(\varphi(\lambda)) = \psi(x^\lambda)$$

is continuous in the same interval. Since:

$$f^0(t, x_1(t)) = f(t, x_1(t)) = \dot{x}_1(t); \quad f^1(t, x_2(t)) = f(t, x_2(t)) = \dot{x}_2(t),$$

from the uniqueness of the solution of problem P) we obtain $x^0 = x_1 \in S_1$, $x^1 = x_2 \in S_2$. Hence:

$$\psi(0) = -\tilde{h}(x_1, S_2) \leq -\beta \quad \text{and} \quad \psi(1) = \tilde{h}(x_2, S_1) \geq \beta.$$

Applying the intermediate value theorem to the continuous function ψ, there exists a $\lambda_\varepsilon \in (0,1)$ such that $\psi(\lambda_\varepsilon) = 0$.

Now let $\{\varepsilon_n\}_{n \in N}$ be a sequence with $\lim_{n \to \infty} \varepsilon_n = \varepsilon$. Set $x_n = x^{\lambda_{\varepsilon_n}}$, consider the sequence $\{x_n\}_{n \in N}$. All the functions x_n are solutions of P), so that $x_n(0) = 0$,

$\forall n \in N$. Let $y_n(t)$ be the function representing the error which is made approximating the derivatives of solutions of C) with $\dot{x}_n(t)$. Then:

$$\dot{x}_n(t) = f(t, x_n(t)) + y_n(t) \quad \text{with} \quad x_n(0) = 0$$

where $\|y_n(t)\| = \|\dot{x}^{\lambda_{\varepsilon_n}}(t) - f(t, x_n(t))\| = \|f^{\lambda_{\varepsilon_n}}(t, x_n(t)) - f(t, x_n(t))\| \leq 4\varepsilon_n$. Hence $\lim_{n\to\infty} \|\dot{x}_n(t) - f(t, x_n(t))\| = 0$.

It follows that the sequence $\{x_n\}_{n\in N}$ admits a subsequence converging uniformly to a solution x of the problem C), that is $x \in S$.

But $\phi(x) = \lim_{n\to\infty} \phi(x_n) = \lim_{n\to\infty} \phi(x^{\lambda_{\varepsilon_n}}) = \lim_{n\to\infty} \psi(\lambda_{\varepsilon_n}) = 0$, a contradiction. Therefore S is a continuum and, by the continuity of the projection $x \to x(t)$, $S(t)$ is a continuum in X.

q.e.d.

Naturally in finite dimensional spaces every continuous function satisfies the condition a) of the theorem, since both sides vanish, so the Peano, Kneser and Hukuhara theorems represent particular cases of the above theorem.

Concerning the statement of Theorem 3.4, observe that all functions of the form $g(x) = kx$ with $k \in \mathbb{R}^+$ are Kamke functions. Therefore we can replace Hypothesis a) with the stronger condition:

$$\alpha(f(I \times A)) \leq k \cdot \alpha(A) \quad \text{for} \quad A \subset D \quad \text{with} \quad k \quad \text{a constant,}$$

preserving the validity of the statement. Consequently, Theorem 3.4 contains previous results due to Szufla [133] and Kubiaczyk [84] who used the hypothesis quoted above; their results contain Ambrosetti's theorem (see [2]) which assumed:

$$\alpha(f(t, A)) \leq k \cdot \alpha(A) \quad \forall t \in I \quad \text{and} \quad \forall A \subset D$$

together with the hypothesis of uniform continuity of f.

It is interesting to note that almost all the results related to Theorem 3.4 differ only in small variations on the Hypothesis a). For example in Szufla [142] a slightly weaker assumption is used:

$$\lim_{d\to 0^+} \alpha(f(I_d \times A)) \leq g(t, \alpha(A))$$

where $I_d \equiv [t, t+d] \cap [0, b]$ and g is a Kamke function.

Rzepecki (see [119] and [120]) instead assumes Hypothesis a) together with uniform continuity of f, and used a function α more general than the Kuratowski measure of noncompactness. In his articles he constructs the solutions of problem C) by means of the method of Euler polygons and the partial ordering induced by a cone.

However, the most important result goes back to 1982 and is due to Szufla [139]. He was inspired by an article of Pianigiani (see [113]) and instead of Hypothesis a), he used the following condition:

$\forall\, \varepsilon > 0$ and $\forall\, A \subset D$ $\exists$ a closed subset $I_\varepsilon \subset I$ with

$\mu(I \setminus I_\varepsilon) < \varepsilon$ such that for each closed set

$$C \subset I_\varepsilon \text{ we have } \alpha\big(f(C \times A)\big) \leq \sup_{t \in C} g\big(t, \alpha(A)\big). \qquad \text{b)}$$

This new assumption holds whenever $f = f_1 + f_2$ with f_1 a compact operator and f_2 satisfying a Kamke condition, that is $\|f_2(t,x) - f_2(t,y)\| \leq g(t, \|x-y\|)$. Thus Szufla's result contains as a particular case a theorem of Krasnoselskii and Krein (see [80]) in which the additive hypothesis is:

$$f = f_1 + f_2 \quad \text{with} \quad f_1 \quad \text{compact and} \quad \|f_2(t,x) - f_2(t,y)\| \leq k(t)\|x - y\|$$

where $k : \mathbb{R} \to \mathbb{R}^+$ is an integrable function.

Szufla's result also includes those of Wazewski [156] and Olech [104], for which they assumed that f satisfies a Kamke condition. Observe that these three papers, written in the fifties and sixties, are the only ones that do not make use of a measure of noncompactness.

In general it seems that Hypothesis a) cannot be replaced by the weaker assumption:

$$\alpha\big(f(t,A)\big) \leq g\big(t, \alpha(A)\big) \quad \forall\, A \subset D. \qquad \text{c)}$$

Nevertheless Monch and von Harten [92] and then Song [127] have recently shown that condition c) is sufficient when X is a w.c.g. space (for example a separable or reflexive space) or $2g$ is a Kamke function.

Finally we remark that in some of the quoted articles only existence is proved. On this point, it is important to observe that each known existence theorem also guarantees the compactness and connectedness of the solution set S (hence S is a continuum), as pointed out by Szufla in [142].

Now let us examine briefly the topological properties of the Peano funnel for the case of a differential equation on a locally convex space endowed with the strong topology.

The main difficulty that we encounter in extending Theorem 3.4 to these spaces is due to the fact that measures of noncompactness are based on the concepts of diameter and ball and therefore are valid only in metric spaces. Thus we need to define a measure of noncompactness for locally convex spaces which reduces to the Hausdorff measure of noncompactness in Banach spaces.

Let X be a quasi-complete locally convex space, P a family of continuous seminorms generating the topology of X and $p \in P$. By a p-bounded set we mean a set $A \subset X$ such that $\sup\limits_{x,y \in A} p(x-y) < \infty$. For each p-bounded set the measure of noncompactness relative to the seminorm p is expressed by:

$$\chi_p(A) = \inf\{\varepsilon > 0 : \exists \text{ a finite set } \{x_1, \ldots, x_n\} \text{ of } X \text{ such that}$$
$$A \subset \{x_1, \ldots, x_n\} + B_p(\varepsilon)\}$$

where $B_p(\varepsilon) = \{x \in X : p(x) \leq \varepsilon\}$ is the "ball" relative to seminorm p.

The characteristics of the function χ_p are exactly the same as for the Hausdorff measure of noncompactness χ_p, except for one property: a set $A \subset X$ is totally bounded if and only if $\chi_p(A) = 0$ for each seminorm $p \in P$. In 1984 Szufla [140] used the functions χ_p to prove a theorem which generalizes previous results obtained in Banach spaces.

Theorem 3.5. *Let X be a quasi-complete locally convex space, $x_0 \in X$, P a family of continuous seminorms in X and $D = \{x \in X : p_i \in P \text{ and } p_i(x - x_0) \leq b \text{ for } i = 1, \ldots, r\}$.*

Suppose that f is a bounded and continuous function from $[0, a] \times X$ into X. Set:

$$M = \sup \{p_i(f(t,x)) : x \in D, \ t \in [0,a] \text{ and } i = 1, \ldots, r\},$$

$$d = \min\left(a, \frac{b}{M}\right) \text{ and } I = [0, d].$$

Then the set of solutions of the problem C) defined on I is a continuum in $C(I, X)$ if one of the following conditions is true $\forall p \in P$ and $\forall A \subset D$:

a) $\chi_p(f(I \times A)) \leq g_p(\chi_p(A))$.

b) $\forall\, p \in P$ the space X is p-separable and we have $\chi_p(f(t,A)) \leq g_p(t, \chi_p(A))$ $\forall\, t \in I$.

c) the function f is uniformly continuous and $\chi_p(f(t,A)) \leq g_p(t, \chi_p(A))$ $\forall\, t \in I$, where g_p is a Kamke function $\forall\, p \in P$.

Proof. We shall only sketch the fundamental lines of the proof. At first Szufla defines a function $\widetilde{f} : I \times X \to X$ equal to f on the convex neighborhood of zero, $D - x_0$. Consequently, proving the existence of solutions of the problem C) is equivalent to finding a solution for the problem with the function $\widetilde{f}$ replacing f.

Set $d_n = \frac{d}{n}$, consider the following functions $x_n : I \to X$:

$$x_n(t) = \begin{cases} x_0 & \text{for } t \in [0, d_n] \\ x_0 + \int_{t_0}^{t-d_n} \widetilde{f}(s, x_n(s))\,ds & \text{for } t \in [d_n, d]. \end{cases}$$

In the rest of the proof, Szufla succeeds in proving the precompactness of the sequence $\{x_n\}_{n \in N}$ by means of the properties of χ_p; an accumulation point x will be a solution of problem C).

Finally, he uses a method very similar to that of Theorem 3.4 to prove the compactness and connectedness of the solution set.

For another discussion of this topic see Dubois and Morales [49].

In the preceding chapter we have seen that in particular spaces equipped with particular topologies, the assumption that f is continuous guarantees the existence of solutions of the problem C). Theorem 2.7 treats the case of $\sigma(X^*, X)$ and $\lambda(X^*, X)$ the infimum and supremum of those topologies for a locally convex space E satisfying the condition $E = (X^*, \tau)$ for a normed barrelled X.

Now we want to study the structure of the solution set under the assumptions of Theorem 2.7, limiting ourselves to the case of Banach spaces endowed with the weak topology. A function $x : [t_0, t_0 + a] \to X$ is said to be a weak solution of C) if:

i) x is weakly differentiable on $[t_0, t_0 + a]$

ii) $x(t_0) = x_0$

iii) $\dot{x}(t) = f(t, x(t))$ $\forall\, t \in [t_0, t_0 + a]$ where $\dot{x}$ denotes the weak derivative of x.

It is known ([130]) that the weak solutions of C) coincide with the solutions of the integral equation:

$$x(t) = x_0 + \int_{t_0}^{t} f(s, x(s))\,ds,$$

where the integral is in the sense of Pettis.

Below, $C_d(I,X)$ denotes the space of weakly continuous functions $x : I \subset \mathbb{R} \to X$ provided with the topology of weak uniform convergence. This topology is determined at each point x_0 by the basis:

$$U_{x_0}(h_1, h_2, \ldots h_n, \varepsilon) = \bigcap_{k=1}^{n} \{x \in C_d(I,x) : \sup_{t \in I} |h_k(x(t) - x_0(t))| < \varepsilon\}$$

with $h_1 \ldots h_n \in X^*$, $\varepsilon > 0$ and $n \in N$.

Szép [130] presents some useful results about weakly continuous functions:

Theorem 3.6. Let $J \equiv [0,d] \subset \mathbb{R}$ and $x : J \to X$ be a weakly continuous function such that $\|x(t)\| \leq M \quad \forall\, t \in J$. Then we have:

$$\left\| \int_p^q x(s)ds \right\| \leq M(p-q) \quad \text{for} \quad 0 \leq p \leq q \leq d.$$

Theorem 3.7. Let $B = B(x_0, r)$ be a ball in X and $f : J \times B \to X$ a weak-weak continuous function. Then $\forall\, \varepsilon > 0$ and $\forall\, h \in X^*$ $\exists$ a weak neighborhood $U \subset X$ of zero such that $\forall\, x, y : J \to B$ weakly continuous functions with $x(t) - y(t) \in U$ and $\forall\, t \in J$, we have:

$$\int_0^d |h(f(s, x(s)) - f(s, y(s)))| ds \leq \varepsilon.$$

Now we can prove the following result of Szufla (see [138]):

Theorem 3.8. Let X be a reflexive Banach space, $I \equiv [0, a]$ and $B = B(x_0, r)$ a ball in X. Assume that $f : I \times B \to X$ is weak-weak continuous with $M = \sup_{(t,x) \in I \times B} \|f(t,x)\|$. Set $d = \min(a, \frac{r}{M})$ and $J \equiv [0, d]$. Then the set of weak solutions of C) is a continuum in $C_d(J, X)$.

Proof. Observe that the constant M exists by the weak compactness of B. Let S_η be the set of functions $x : J \to X$ such that:

i) $x(0) = x_0$ and $\|x(t) - x(s)\| \leq M \cdot |t - s| \quad \forall\, t, s \in J$

ii) $\sup_{t \in J} \|x(t) - x_0 - \int_0^t f(s, x(s)) ds\| < \eta$.

Let $\varepsilon \in \mathbb{R}$ with $0 < \varepsilon < \beta = \min(d, \frac{\eta}{M})$. Define $v_\varepsilon : J \to X$ as:

$$v_\varepsilon(t) = \begin{cases} x_0 & \text{for } t \in [0, \varepsilon] \\ x_0 + \int_0^{t-\varepsilon} f(s, v_\varepsilon(s)) ds & \text{for } t \in [\varepsilon, d]. \end{cases}$$

Applying Theorem 3.6 we have:

$$\|v_\varepsilon(t) - x_0\| = \left\| \int_0^{t-\varepsilon} f(s, v_\varepsilon(s)) ds \right\| \leq M(t - \varepsilon) \quad \text{for } t \in [\varepsilon, d]$$

$$\|v_\varepsilon(t) - v_\varepsilon(s)\| \leq \left\| \int_{s-\varepsilon}^{t-\varepsilon} f(z, v_\varepsilon(z)) dz \right\| \leq M \cdot |t - s| \quad \text{for } t \in J$$

$$\left\| v_\varepsilon(t) - x_0 - \int_0^t f(s, v_\varepsilon(s)) ds \right\| = \left\| \int_0^t f(s, x_0) ds \right\| \leq M\varepsilon < \eta \quad \text{for } t \in [0, \varepsilon]$$

$$\left\| v_\varepsilon(t) - x_0 - \int_0^t f(s, v_\varepsilon(s)) ds \right\| = \left\| \int_{t-\varepsilon}^t f(s, v_\varepsilon(s)) ds \right\| \leq M\varepsilon < \eta \quad \text{for } t \in [\varepsilon, d].$$

Consequently $v_\varepsilon \in S_\eta \quad \forall \varepsilon \in (0, \beta)$.

Let δ be a constant such that $0 < \varepsilon < \delta < \beta$. Then:

$$\|v_\delta(t) - v_\varepsilon(t)\| = \|x_0 - x_0\| = 0 \quad \text{for } t \in [0, \varepsilon]$$

$$\|v_\delta(t) - v_\varepsilon(t)\| = \left\| \int_0^{t-\varepsilon} f(s, v_\varepsilon(s)) ds \right\|$$
$$\leq M(t - \varepsilon) \leq M(\delta - \varepsilon) \quad \text{for } t \in [\varepsilon, \delta].$$

So $\forall h \in X^*$ with $\|h\| \leq 1$ we have:

$$|h(v_\delta(t) - v_\varepsilon(t))| \leq \|h\| \cdot \|v_\delta(t) - v_\varepsilon(t)\| \leq M(\delta - \varepsilon) \quad \text{for } t \in [\varepsilon, \delta].$$

For $t \in [\delta, d]$ we have

$$|h(v_\delta(t) - v_\varepsilon(t))|$$

$$= \left| h\left(\int_0^{t-\delta} f(s, v_\delta(s)) ds - \int_0^{t-\varepsilon} f(s, v_\varepsilon(s)) ds \right) \right|$$

$$= \left| h\left(\int_0^{t-\varepsilon} f(s, v_\delta(s)) ds - \int_{t-\delta}^{t-\varepsilon} f(s, v_\delta(s)) ds - \int_0^{t-\varepsilon} f(s, v_\varepsilon(s)) ds \right) \right|$$

$$\le \left| h \left(\int_{t-\delta}^{t-\varepsilon} f(s, v_\delta(s)) ds \right) \right| + \left| h \left(\int_0^{t-\varepsilon} f(s, v_\delta(s)) - f(s, v_\varepsilon(s)) ds \right) \right|$$

$$\le M(\delta - \varepsilon) + \int_0^{t-\varepsilon} \left| h(f(s, v_\delta(s)) - f(s, v_\varepsilon(s))) \right| ds.$$

Then, taking ε and δ such that $\varepsilon > 0$ and $\delta < \beta$, the following inequalities are true:

a) $\|v_\delta(t) - v_\varepsilon(t)\| \le M \cdot |\delta - \varepsilon|$ for $t \in [0, \varepsilon]$

b) $|h(v_\delta(t) - v_\varepsilon(t))| \le M \cdot |\delta - \varepsilon| + \int_0^{t-\varepsilon} |h(f(s, v_\delta(s)) - f(s, v_\varepsilon(s)))| ds$ for $t \in [\varepsilon, d]$.

From a) we have:

$$\lim_{\delta \to \varepsilon} v_\delta(t) = v_\varepsilon(t) \quad \text{strongly, uniformly on} \quad [0, \varepsilon].$$

Naturally on the same interval weak uniform convergence also holds.

Taking $t \in [\varepsilon, 2\varepsilon]$, from b) we have:

$$|h(v_\delta(t) - v_\varepsilon(t))| \le M \cdot |\delta - \varepsilon| + \int_0^\varepsilon |h(f(s, v_\delta(s)) - f(s, v_\varepsilon(s)))| ds.$$

Notice that the variable s in the integral varies over $[0, \varepsilon]$ and in this interval v_δ converges in norm to v_ε as $\delta \to \varepsilon$. Therefore, if U is some weak neighborhood of zero in X, from some δ on we shall have:

$$v_\delta(t) - v_\varepsilon(t) \in U$$

and by Theorem 3.7 we obtain:

$$\int_0^\varepsilon |h(f(s, v_\delta(s)) - f(s, v_\varepsilon(s)))| ds \to 0 \quad \text{as} \quad \delta \to \varepsilon.$$

Hence:

$$\lim_{\delta \to \varepsilon} |h(v_\delta(t) - v_\varepsilon(t))| = 0 \quad \text{on} \quad [\varepsilon, 2\varepsilon] \quad \forall \, \|h\| \le 1$$

$$\lim_{\delta \to \varepsilon} v_\delta(t) = v_\varepsilon(t) \quad \text{weakly, uniformly on} \quad [\varepsilon, 2\varepsilon] \cup [0, \varepsilon] \equiv [0, 2\varepsilon].$$

By repeating these arguments we see that $v_\delta(t)$ converges uniformly to $v_\varepsilon(t)$ on $[0, n\varepsilon]$ for $n = 1, 2, \ldots, [\frac{d}{\varepsilon}]$, which proves that:

$$\lim_{\delta \to \varepsilon} v_\delta(t) = v_\varepsilon(t) \quad \text{weakly, uniformly on} \quad J.$$

Consequently, the function $\phi : (0, \beta) \subset \mathbb{R} \to C_d(J, X)$ defined as:

$$\phi(\varepsilon) = v_\varepsilon$$

is continuous on $(0, \beta)$ and the set $V = \{v_\varepsilon : \varepsilon \in (0, \beta)\} \equiv \phi((0, \beta))$ will be connected in $C_d(J, X)$.

Now let $x \in S_\eta$. From property ii) of S_η we deduce that for $\varepsilon \in (0, \beta)$ sufficiently small,

$$\sup_{t \in J} \left\| x(t) - x_0 - \int_0^t f(s, x(s))\,ds \right\| + M\varepsilon < \eta.$$

Fix such an ε. For a given $p \in [0, d]$, let $y_p : J \to X$ be the following function:

$$y_p(t) = \begin{cases} x(t) & \text{for} \quad 0 \leq t \leq p \\ x(p) & \text{for} \quad p \leq t \leq \min(d, p + \varepsilon) \\ x(p) + \int_p^{t-\varepsilon} f(s, y_p(s))\,ds & \text{for} \quad \min(d, p + \varepsilon) \leq t \leq d. \end{cases}$$

From Theorem 3.6 we obtain the inequalities:

$$\|y_p(t) - x_0\| \leq \|x(p) - x_0\| + \left\| \int_p^{t-\varepsilon} f(s, y_p(s))\,ds \right\|$$

$$\leq Mp + M(t - \varepsilon - p) = M(t - \varepsilon)$$

$$\leq M(d - \varepsilon) \leq Md \leq r \qquad \text{for} \quad \min(d, p + \varepsilon) \leq t \leq d$$

$$\|y_p(t) - y_p(s)\| \leq M \cdot |t - s| \qquad \text{for} \quad t, s \in J$$

$$\left\| y_p(t) - x_0 - \int_0^t f(s, y_p(s)) ds \right\|$$

$$= \left\| x(p) - x_0 - \int_0^p f(s, y_p(s)) ds - \int_p^t f(s, y_p(s)) ds \right\|$$

$$\leq \left\| x(p) - x_0 - \int_0^p f(s, x(s)) ds \right\| + \left\| \int_p^t f(s, y_p(s)) ds \right\|$$

$$\leq \sup_{t \in J} \left\| x(t) - x_0 - \int_0^t f(s, x(s)) ds \right\| + M(t - p) < \eta$$

for $p \leq t \leq \min(d, p + \varepsilon)$

and

$$\left\| y_p(t) - x_0 - \int_0^t f(s, y_p(s)) ds \right\|$$

$$\leq \left\| x(p) - x_0 - \int_0^p f(s, x(s)) ds \right\| + \left\| \int_{t-\varepsilon}^t f(s, y_p(s)) ds \right\|$$

$$\leq \sup_{t \in J} \left\| x(t) - x_0 - \int_0^t f(s, x(s)) ds \right\| + M\varepsilon < \eta$$

for $\min(d, p + \varepsilon) \leq t \leq d$.

Therefore $y_p \in S_\eta \quad \forall p \in J$.

Choose a number q such that $0 \leq p \leq q \leq d$ and $q \in [p, p + \varepsilon]$. Then:

$$\|y_q(t) - y_p(t)\| = \|x(t) - x(t)\| = 0 \quad \text{for } t \in [0, p]$$

$$\|y_q(t) - y_p(t)\| = \|x(t) - x(p)\| \quad \text{for } t \in [p, q]$$

$$\|y_q(t) - y_p(t)\| = \|x(q) - x(p)\| \quad \text{for } q \leq t \leq \min(d, p + \varepsilon).$$

Three cases are possible regarding the position of the point $d - \varepsilon$:

a) $p \leq q \leq d - \varepsilon$ b) $p \leq d - \varepsilon \leq q$ c) $d - \varepsilon \leq p \leq q$.

We consider the first case:

$$\|y_q(t) - y_p(t)\| = \left\| x(q) - x(p) - \int_p^{t-\varepsilon} f(s, y_p(x)) ds \right\|$$

$$\leq \|x(q) - x(p)\| + \left\| \int_p^{t-\varepsilon} f(s, y_p(s)) ds \right\|$$

$$\leq M(q-p) + M(t-\varepsilon-p) \leq M(q-p) + M(q+\varepsilon-\varepsilon-p)$$

$$= 2M(q-p) \quad \text{for } t \in [p+\varepsilon, q+\varepsilon].$$

Using Theorem 3.6, for $t \in [q+\varepsilon, d]$ we have:

$$|h(y_q(t) - y_p(t))| = \left| h\left(x(q) + \int_q^{t-\varepsilon} f(s, y_q(s)) ds - x(p) - \int_p^{t-\varepsilon} f(s, y_p(s)) ds \right) \right|$$

$$\leq \|x(q) - x(p)\| + \left\| \int_p^q f(s, y_p(s)) ds \right\|$$

$$+ \left| \int_q^{t-\varepsilon} h(f(s, y_q(s)) ds - f(s, y_p(s))) ds \right|$$

$$\leq 2M(q-p) + \int_q^{t-\varepsilon} |h(f(s, y_q(s)) ds - f(s, y_p(s)))| \, ds$$

with $h \in X^*$, $\|h\| \leq 1$.

Now we pass to case b):

$$\|y_q(t) - y_p(t)\| = \left\| x(q) - x(p) - \int_0^{t-\varepsilon} f(s, y_p(s)) ds \right\|$$

$$\leq \|x(q) - x(p)\| + \left\| \int_p^{t-\varepsilon} f(s, y_p(s)) ds \right\|$$

$$\leq 2M(q-p) \quad \text{for } t \in [p+\varepsilon, d].$$

Finally in case c) we have:

$$\|y_q(t) - y_p(t)\| = \|x(q) - x(p)\| \leq M(q-p) \quad \text{for } t \in [q, d].$$

So, given two numbers p and q, with $0 \leq p$, $q \leq d$ and $|q-p| \leq \varepsilon$, the following conditions are satisfied:

*) $\|y_q(t) - y_p(t)\| \leq 2M \cdot |p - q|$ for $t \in [0, \varepsilon]$.

**) $|h(y_q(t) - y_p(t))| \leq 2M \cdot |p - q| + \int_q^{t-\varepsilon} |h(f(s, y_q(s))ds - f(s, y_p(s)))| ds$ for $t \in [\varepsilon, d]$.

By *) we obtain:

$$\lim_{p \to q} y_p(t) = y_q(t) \quad \text{strongly, uniformly on} \quad [0, \varepsilon].$$

From Theorem 3.7, with a procedure identical to one used for $v_\varepsilon(t)$, we obtain

$$\lim_{p \to q} y_p(t) = y_q(t) \quad \text{weakly, uniformly on} \quad [0, 2\varepsilon].$$

Repeating this proof, we can see that y_p tends weakly to y_q, uniformly on $[0, n\varepsilon]$ for $n = 1, 2, \ldots [\frac{d}{\varepsilon}]$. Hence:

$$\lim_{p \to q} y_p(t) = y_q(t) \quad \text{weakly, uniformly on} \quad J.$$

So the function $\varphi : J \subset \mathbb{R} \to C_d(J, X)$ defined as $\varphi(p) = y_p$ is continuous on J and the set $T_x = \{y_p : p \in J\} = \varphi(J)$ is connected in $C_d(J, X)$.

Since by definition $y_0(t) = v_\varepsilon(t) \in V \cap T_x$ for $x \in S_\eta$, the connectedness of V and T_x implies that the set $V \cup T_x$ will be connected in $C_d(J, X)$, thus also the set $W = \bigcup_{x \in S_\eta} V \cup T_x$. Moreover $\forall\, x \in S_\eta$ we have $x = y_d \in T_x \subset W$ so $S_\eta \subset W$. But we have proved $T_x \subset S_\eta$ and $V \subset S_\eta$, so that $W \subset S_\eta$. Therefore the set $S_\eta \equiv W$ is a connected subset of $C_d(J, X)$.

Now let $x \in \overline{S}_\eta$ and $t \in J$. By a corollary of the Hahn-Banach theorem there exists in X^* a functional h with $\|h\| = 1$ such that:

$$h\left(x(t) - x_0 - \int_0^t f(s, x(s)) ds\right) = \left\|x(t) - x_0 - \int_0^t f(s, x(s)) ds\right\|.$$

From Theorem 3.7 we conclude that $\forall\, \varepsilon > 0$ $\exists$ a weak neighborhood U of zero such that for each weakly continuous function $y : J \to B$ with $x(s) - y(s) \in U$ $\forall\, s \in J$, we have:

$$|h(x(t) - y(t))| \leq \varepsilon \quad \text{and} \quad \int_0^t |h(f(s, x(s)) - f(s, y(s)))| ds \leq \varepsilon.$$

As $x \in \overline{S}_\eta$, by the definition of a closed set there exists $y \in S_\eta$ such that $x(s) - y(s) \in U \quad \forall s \in J$. Therefore the following inequality holds:

$$h\left(x(t) - x_0 - \int_0^t f(s, x(s))ds\right)$$

$$= h\left(x(t) - x_0 - \int_0^t f(s, x(s))ds + y(t) + \int_0^t h(f(s, y(s)))ds\right)$$

$$- y(t) - \int_0^t h(f(s, y(s)))ds$$

$$\leq |h(x(t) - y(t))| + \left|\int_0^t h(f(s, y(s))) - f(s, x(s)))ds\right| +$$

$$+ \left\|y(t) - x_0 - \int_0^t f(s, y(s))ds\right\|$$

$$\leq \varepsilon + \varepsilon + \eta = 2\varepsilon + \eta,$$

thus

$$\left\|x(t) - x_0 - \int_0^t f(s, x(s))ds\right\| \leq 2\varepsilon + \eta$$

and, since ε is arbitrary, we have:

$$\left\|x(t) - x_0 - \int_0^t f(s, x(s))ds\right\| \leq \eta.$$

Consequently $x \in S_\eta$ and the set S_η is closed in $C_d(J, X)$.

From property i) it follows that S_η is equicontinuous. Since $S_\eta(t) = \{x(t) : x \in S_\eta\} \subset B$ and the ball B is weakly compact in the reflexive space X (see [47]), the Ascoli-Arzelà theorem implies that S_η is compact in $C_d(J, X)$.

Set $W_n \equiv S_{1/n} \quad \forall n \in N$. Then $\{W_n\}_{n \in N}$ is a decreasing sequence of continua in $C_d(J, X)$ whose intersection $S = \bigcap_{n=1}^\infty W_n$ is again a continuum. Finally property ii) implies that S and the set of weak solutions of the problem C) coincide.

q.e.d.

Observe that Theorem 3.8 remains true if we consider a locally convex space $X = (E^*, \tau)$ where E is normed barrelled and τ is a topology included between

$\sigma(X, X^*)$ and $\lambda(X, X^*)$. In fact in the proof we can replace σ with λ since, according to the Banach-Dieudonné theorem, the topology of precompact convergence is the strongest of those topologies equivalent to the weak one on equicontinuous subsets of X. We note that these are the unique known cases in which the assumption that f is continuous is sufficient to assure the continuity of the solution set, as in finite dimensional spaces.

Clearly the preceding theorem fails without the hypothesis of reflexivity on the space X. In the case that X is not reflexive we must add a new condition which is usually expressed by means of the measure of noncompactness β_d developed by De Blasi [34] and defined $\forall\, A \subset X$ as:

$$\beta_d(A) = \inf\{\varepsilon > 0 : \exists \text{ a weakly compact set } C \subset X \text{ such that } A \subset C + \varepsilon B_1\}$$

where B_1 is the unit ball in X. Obviously β_d vanishes on weakly precompact sets in X. The first study of the set of weak solutions in nonreflexive Banach spaces was by Kubiaczyk [84] in 1980, followed two years later by Rzepecki [118]. In their articles both of them set as supplementary condition:

$$\beta_d(f(I \times A)) \leq k \cdot \beta_d(A) \quad \forall\, A \subset B(x_0, r).$$

This assumption was weakened by Kubiaczyk-Szufla [82] as follows:

$$\beta_d(f(I \times A)) \leq h(\beta_d(A)) \quad \forall\, A \subset B(x_0, r)$$

where $h : \mathbb{R}^+ \to \mathbb{R}$ is a Kamke function.

The most general result that is known at present is due to Papageorgiou [105], in 1986. He assumed

$$\lim_{\tau \to 0^+} \beta_d(f(I_\tau \times A)) \leq h(t, \beta_d(A))$$

almost everywhere with $I_\tau \equiv [t_0, t_0 + \tau]$.

4. Aronszajn Type Theorems

It has been known since 1928 that in finite-dimensional spaces the solution set of the problem C) with continuous right hand side is a continuum in the space of continuous functions. Later on, Aronszajn [5] succeeded in defining a new topological concept suitable for more precisely describing the structure of this set. In his article, written in 1940 but published two years later because of the war, he introduced the notion of an R_δ-set as the intersection of a decreasing sequence of compact absolute retracts and showed that in finite dimensional spaces the Peano funnel is homeomorphic to a compact R_δ (always assuming continuous right hand side). Notice that an R_δ is an acyclic set (see [88]), that is it has the cohomology of a single point. Therefore without a Lipschitz assumption the solution set S of a differential equation may not consist of an unique element, but, from the point of view of algebraic topology, S is equivalent to a point, in the sense that it has the same cohomology. Subsequently Aronszajn's result was improved for finite dimensional spaces by several authors, such as Browder and Gupta [17], Vidossich ([151] and [152]), Lasry and Robert [87] and Bielawski and Pruszko [12]. The aim of this chapter will be to extend these results to Banach spaces.

To this end we must define some concepts that will be useful later.

Let X_1 be a closed subset of a Banach space X and $F : X_1 \to X$ an operator. We shall say that F satisfies the **Palais-Smale condition** on X_1 if from each sequence $\{x_n\}_{n \in N}$ in X_1 such that $x_n - F(x_n) = 0$, we can extract a convergent subsequence. We shall say that F is **0-closed** if for every closed set $V \subset X_1$, the inclusion $0 \in \overline{F(V)}$ implies $0 \in F(V)$.

These two notions are connected by the following theorem due to Dubois and Morales [50]:

Theorem 4.1. *Let X be a Banach space and X_1 a closed subset. Suppose that the operator $F : X_1 \to X$ is continuous on X_1 and the set S of fixed points of F is nonempty and compact. Then the following conditions are equivalent:*

a) *F satisfies the Palais-Smale condition;*
b) *$\lim_{n \to \infty} \alpha(S_n) = 0$ where $S_n = \{x \in X_1 : \|x - F(x)\| \leq \frac{1}{n}\}$*
c) *the operator $T = I - F$ is 0-closed.*

Notice that in fact conditions a) and b) are equivalent without any assumption on the fixed points of F, while condition a) is stronger than c) and in fact can be deduced from c) only if the hypotheses of Theorem 4.1 hold.

Starting from Aronszajn's definition of an R_δ-set, many equivalent characterizations have been obtained. One such is due to Browder and Gupta [17]:

Theorem 4.2. *Let X be a metric space and $\{R_n\}_{n \in N}$ a sequence of compact absolute retracts of X. Suppose that S is a subset of X satisfying the following conditions:*
*) $S \subseteq R_n \quad \forall n \in N$
**) $\lim_{n \to \infty} R_n = S$
***) *for each neighborhood V of S $\exists$ a subsequence $\{R_{n_k}\}_{k \in N}$ of $\{R_n\}_{n \in N}$ such that $R_{n_k} \subseteq V \quad \forall k \in N$.*
Then S is a compact R_δ of X.

The next theorem will show some sufficient conditions in order that the set of fixed points of a continuous operator be a compact R_δ. A similar result, but with stronger and more complex assumptions, was proved by Vidossich [153] and it is stated without proof in Szufla [134].

Theorem 4.3. *Let X be a Banach space and $T : X \to X$ a continuous and 0-closed mapping. Suppose that T is the uniform limit of a sequence $\{T_n\}_{n \in N}$ of homoeomorphisms from X into X. Then we have:*
a) $S = T^{-1}(0)$ *is nonempty*
b) *If S is compact, then S is an R_δ set of X.*

Proof. a) Obviously the homeomorphisms T_n are surjective; so $\forall n \in N \ \exists \ x_n \in X$ such that $T_n(x_n) = 0$. Hence:

$$0 \leq \|T(x_n)\| = \|T_n(x_n) - T(x_n)\| \leq \|T_n - T\| \equiv \sup_{x \in X} \|T_n(x) - T(x)\|, \quad \forall n \in N.$$

By hypothesis we have $\lim_{n \to \infty} \|T_n - T\| = 0$ and so

$$\lim_{n \to \infty} T(x_n) = 0.$$

In other words $0 \in \overline{T(X)}$ and, since T is 0-closed, we have $0 \in T(X)$, so $S = T^{-1}(0) \neq \emptyset$. This part of the theorem does not require that X be complete, nor need T be continuous, and the operators T_n need only be surjective.

b) Since by hypothesis $\|T_n - T\| \to 0$ as $n \to \infty$, we can assume that:

$$\sup_{x \in X} \|T_n(x) - T\| < \frac{1}{n} \quad \forall\, n \in N.$$

As T_n is a homeomorphism and S is compact, the set $T_n(S)$ is compact. Given a point $y \in T_n(S)$, $\exists\, x \in S$ such that $T_n(x) = y$. Therefore:

$$\|y\| = \|T_n(x)\| \le \|T_n(x) - T(x)\| + \|T(x)\| < \frac{1}{n}.$$

Hence $y \in B(0, \frac{1}{n})$ and $T_n(S) \subset B(0, \frac{1}{n})\ \forall\, n \in N$.

Since each convex, closed and bounded subset of a normed space is an absolute retract, the sets $Q_n = \overline{co}\,\{T_n(S)\}$ will be compact absolute retracts (by the completeness of X) and the same is true for the sets $R_n = T_n^{-1}(Q_n)$.

We want to show that S and $\{R_n\}_{n \in N}$ satisfy the hypotheses of Theorem 4.2. First, we have:

$$S \subseteq T_n^{-1}(T_n(S)) \subseteq T_n^{-1}(\overline{co}\,\{T_n(S)\}) = R_n$$

so that the condition $*$) is satisfied. Moreover $\forall\, n \in N$ we have $S \subseteq \bigcap_{k=n}^{\infty} R_k$. Then:

$$S \subseteq \bigcup_{n=1}^{\infty} \bigcap_{k=n}^{\infty} R_k = \liminf R_n.$$

Now let $x \in \limsup R_n = \bigcap_{n=1}^{\infty} \bigcup_{k=n}^{\infty} R_k$. As $x \in \bigcup_{k=n}^{\infty} R_k\ \forall\, n \in N$, it is possible to construct a subsequence $\{R_{n_k}\}_{k \in N}$ of $\{R_n\}_{n \in N}$ such that $x \in R_{n_k}\ \forall\, k \in N$.

Applying T_{n_k} to both the sides, we obtain $T_{n_k}(x) \in T_{n_k}(R_{n_k}) = Q_{n_k} \subset B(0, \frac{1}{n_k})$. So the following relations are true:

$$\|T_{n_k}(x)\| \le \frac{1}{n_k} \quad \text{and} \quad \|T_{n_k}(x) - T(x)\| \le \frac{1}{n_k}$$

which implies $T(x) = 0$ and $x \in S$. Consequently, $\limsup R_n \subseteq S \subseteq \liminf R_n$, that is $S = \lim_{n \to \infty} R_n$. Finally it remains to verify the condition $***$) of Theorem 4.2.

Let V be an open neighborhood of S. Suppose that there exists an integer k such that $\forall\, n \ge k$ we have $R_n \not\subset V$, that is to say $\exists\, x_n \in R_n \setminus V$.

Set $W = \overline{\{x_n : n \geq k\}}$. Reasoning as at the beginning of part b) of this proof, we obtain:

$$\|T_n(x_n)\| \leq \frac{2}{n} \quad \text{and} \quad \lim_{n \to \infty} T(x_n) = 0.$$

In other words $0 \in \overline{T(W)}$ and, since T is 0-closed, we have $0 \in T(W)$. Consequently there exists $x_0 \in W$ such that $T(x_0) = 0$ and $x_0 \in S$. As V is open, we see that:

$$W \subset \bigcup_{n=1}^{\infty} (R_n \setminus V) \subset \overline{X \setminus V} = X \setminus V.$$

Hence $x_0 \in W \cap S \subseteq (X \setminus V) \cap S = (X \cap S) \setminus (V \cap S) = S \setminus V = \emptyset$, a contradiction.

q.e.d.

The above theorem is in fact only useful if the function T is the uniform limit of a sequence of homeomorphisms. Our aim will be to find new conditions, easy to verify, which assure this property when T is a mapping from $C(X,X)$ into X. To this end we need a result of Vidossich [152] from 1971, later simplified by Dubois and Morales [50]:

Theorem 4.4. *Let Y be a Banach space, K a bounded and convex subset of a normed linear space X and $Z = C(K,Y)$. Assume that $F : Z \to Z$ is a continuous mapping satisfying the following conditions:*
  i) $\exists\, x_0,\, x_1 \in K$ *such that* $F(f)(x_0) = x_1 \quad \forall\, f \in Z$
  ii) $\forall\, \varepsilon > 0$ *and* $\forall\, f,\, g \in Z$ *with* $f|_{K_\varepsilon} = g|_{K_\varepsilon}$ *we have* $F(f)|_{K_\varepsilon} = F(g)|_{K_\varepsilon}$ *where* $K_\varepsilon = K \cap B(x_0, \varepsilon)$
  iii) $F(Z)$ *is a uniformly equicontinuous set of mappings from K into Y.*
Then there exists a sequence $\{T_n\}_{n \in \mathbb{N}}$ of homeomorphisms from Z into Z which converges uniformly to $T = I - F$.

Proof. We first note that hypothesis iii) means: $\forall\, \varepsilon > 0\, \exists\, \delta > 0$ such that $\forall\, y_1, y_2 \in K$ with $\|y_1 - y_2\| \leq \delta$ we have $\sup_{f \in Z} \|F(f)(y_1) - F(f)(y_2)\| \leq \varepsilon$. We shall say that a sequence $\{T_n\}_{n \in \mathbb{N}}$ converges to $T \in Z$ if:

$$\lim_{n \to \infty} \{\sup_{f \in Z} \sup_{x \in K} \|T_n(f)(x) - T(f)(x)\|\} = 0.$$

Since K is bounded, $\forall\, n \in N$ we can find a number $k_n \in N$ such that $K \subseteq B(x_0, \frac{k_n}{n})$. Set:

$$C_i = \left\{ x \in K : \frac{i-1}{n} \leq \|x - x_0\| \leq \frac{i}{n} \right\} \quad \text{for} \quad i = 1, 2, \ldots, k_n.$$

The sets C_i divide K into concentric rings. Obviously $K = \bigcup_{i=1}^{k_n} C_i$.

Consider the function $r_n : K \setminus \{x_0\} \to X$ defined as:

$$r_n(x) = \left(1 - \frac{1}{n\|x - x_0\|}\right) x + \frac{1}{n\|x - x_0\|} x_0.$$

Let $x \in K$ with $\|x - x_0\| \geq \frac{1}{n}$ and $\alpha = \frac{1}{n\|x - x_0\|}$. Then we have $0 \leq \alpha \leq 1$ and from the convexity of K we obtain $r_n(x) = (1-\alpha)x + \alpha x_0 \in K$. Hence:

P_0) $\forall\, n \in N$ and $\forall\, x \in K$ with $\|x - x_0\| \geq \frac{1}{n}$ we have $r_n(x) \in K$.

Moreover, r_n enjoys other properties:

P_1) $\|r_n(x) - x\| = \left\|\frac{x - x_0}{n\|x - x_0\|}\right\| = 1/n \quad \forall\, x \in K \setminus C_1$

P_2) $r_n(C_1 \cap C_2) = \{x_0\}$.

P_3) $r_n(C_i) \subseteq C_{i-1}$ for $i = 2, \ldots, k_n$.

To prove P_2), note that $C_1 \cap C_2 = \{x \in K : \|x - x_0\| = \frac{1}{n}\}$ and, if $x \in C_1 \cap C_2$, we have $r_n(x) = x_0$

In order to prove P_3), let $x \in C_i$. Then we have $\frac{i-1}{n} \leq \|x - x_0\| \leq \frac{i}{n}$ and:

$$\|r_n(x) - x_0\| = \left\| x - \frac{x - x_0}{n\|x - x_0\|} - x_0 \right\| = \|x - x_0\| \cdot \left|1 - \frac{1}{n\|x - x_0\|}\right|$$

$$= \|x - x_0\| \cdot \left|\frac{n\|x - x_0\| - 1}{n\|x - x_0\|}\right| = \frac{|n\|x - x_0\| - 1|}{n}$$

$$= \left| \|x - x_0\| - \frac{1}{n} \right| \leq \frac{i}{n} - \frac{1}{n} = \frac{i-1}{n}$$

$$\|r_n(x) - x_0\| = \left| \|x - x_0\| - \frac{1}{n} \right| \geq \frac{i-1}{n} - \frac{1}{n} = \frac{i-2}{n}.$$

Therefore $\frac{i-2}{n} \leq \|r_n(x) - x_0\| \leq \frac{i-1}{n}$, that is $r_n(x) \in C_{i-1}$.

$\forall n \in N$ and $\forall f \in C(K,Y)$ set:

$$F_n(f)(x) = \begin{cases} x_1 & \text{if } x \in C_1 \\ F(f)(r_n(x)) & \text{if } x \in K \setminus C_1. \end{cases}$$

The function $F_n(f) : K \to Y$ is constant and continuous on C_1; the same happens in $K \setminus C_1$, since $F_n(f)$ is the composition of the continuous functions $r_n(x)$ and $F(f)$. Finally, using property P_2) and the fact that C_1 is closed, for $x \in \partial C_1 = \overline{C_1} \cap \overline{K \setminus C_1} = C_1 \cap \bigcup_{i=1}^{\infty} C_i = C_1 \cap C_2$ we obtain:

$$F_n(f)(r_n(x)) \in F_n(f)(r_n(C_1 \cap C_2)) = F_n(f)(\{x_0\}) = x_1.$$

So the function $F_n(f)$ is continuous on all of K, that is $F_n(f) \in Z$.

If $f, g \in Z$, from property P_0) we obtain:

$\|F_n(f) - F_n(g)\|$

$= \sup_{x \in K} \|F_n(f)(x) - F_n(g)(x)\|$

$= \max \left\{ \sup_{x \in C_1} \|x_1 - x_1\|, \sup_{x \in K \setminus C_1} \|F(f)(r_n(x)) - F(g)(r_n(x))\| \right\}$

$= \sup_{\|x-x_0\| \geq \frac{1}{n}} \|F(f)(r_n(x)) - F(g)(r_n(x))\|$

$\leq \|F(f) - F(g)\|.$

Thus F_n is a continuous mapping from Z into Z. Given $f \in Z$, Hypotheses i) and ii) imply:

$\|F_n(f) - F(f)\|$

$= \max \left\{ \sup_{\|x-x_0\| \leq \frac{1}{n}} \|x_1 - F(f)(x)\|, \sup_{\|x-x_0\| \geq \frac{1}{n}} \|F(f)(r_n(x)) - F(f)(x)\| \right\}$

$= \sup_{\|x-x_0\| \geq \frac{1}{n}} \|F(f)(r_n(x)) - F(f)(x)\| + o(1) \quad \text{as} \quad n \to \infty.$

But from property P_1) and the uniform continuity of $F(f)$ it follows that:

$$\|F(f)(r_n(x)) - F(f)(x)\| \to 0 \quad \text{as} \quad n \to \infty \quad \text{and}$$

$$\lim_{n \to \infty} F_n = F \quad \text{uniformly on} \quad Z.$$

Now set $T_n = I - F_n$. Naturally $T_n \to I - F = T$. We want to show that each T_n is homeomorphism from Z into Z.

The continuity of F_n implies that T_n is continuous on Z. We check whether T_n is injective. Let f and g be such that $T_n(f) = T_n(g)$. Then $\forall\, x \in C_1$ we have:

$$f(x) - x_1 = g(x) - x_1 \quad \text{and} \quad f(x) = g(x).$$

In other words f and g coincide on C_1. From hypothesis ii) we have:

$$F(f)|_{C_1} = F(g)|_{C_1}$$

and, if $x \in C_2$, using property P_3), we have $r_n(x) \in C_1$. Therefore:

$$F(f)(r_n(x))|_{C_2} = F(g)(r_n(x))|_{C_2}.$$

From this we obtain:

$$F_n(f)(x)|_{C_2} = F_n(g)(x)|_{C_2} \quad \text{thus also} \quad T_n(f)(x)|_{C_2} = T_n(g)(x)|_{C_2}$$

thus f and g coincide on $C_1 \cup C_2$. By repeating the same arguments, we deduce that f and g coincide on K.

Let us show that T_n is surjective on Z. Given $g \in Z$, set:

$$f_1(x) = g(x) + x_1 \quad \text{with } x \in K$$

$$f_2(x) = \begin{cases} f_1(x) & \text{for } x \in C_1 \\ g(x) + F(f_1)(r_n(x)) & \text{for } x \in K \setminus C_1. \end{cases}$$

The function f_2 is continuous on C_1 and on $K \setminus C_1$ as it is the composition of continuous functions. From property P_2) and Hypothesis i) it follows that f_2 is continuous on $\partial C_1 = C_1 \cap C_2$ and $f_2 \in Z$.

Moreover,

$$f_2(x) = g(x) + \begin{cases} x_1 & \text{for } x \in C_1, \\ F(f_1)(r_n(x)) & \text{for } x \in K \setminus C_1, \end{cases}$$

that is $f_2(x) = g(x) + F_n(f_1)(x)$ on K.

Since by definition f_1 and f_2 coincide on C_1, with a similar procedure to the one just used for the proof of injectivity we can see that f_1 and f_2 coincide on $C_1 \cup C_2$. In general for $i = 1, 2, \ldots, k_n - 1$ we can define:

$$f_{i+1}(x) = \begin{cases} f_i(x) & \text{for } x \in C_1, \\ g(x) + F(f_i)(r_n(x)) & \text{for } x \in K \setminus C_1. \end{cases}$$

$$= \begin{cases} f_i(x) & \text{for } x \in C_1 \cup C_2 \cup \cdots \cup C_{i+1}, \\ g(x) + F(f_i)(r_n(x)) & \text{otherwise}. \end{cases}$$

So $f_{i+1}(x) = g(x) + F_n(f_i)(x)$. We have constructed a sequence $\{f_i\}_{i \in \mathbb{N}}$ of continuous functions on K. Denoting the function f_{k_n} by f, we see that $f \in Z$ and:

$$f \equiv f_{k_n} = f_{(k_n-1)+1} \equiv f_{k_n-1} \quad \text{on} \quad \bigcup_{i=1}^{k_n} C_i = K$$

$$f(x) = f_{k_n}(x) = g(x) + F_n(f_{k_n-1})(x) \quad \text{on} \quad K$$

$$f(x) = g(x) + F_n(f)(x) \quad \text{or} \quad (I - F_n)(f)(x) = g(x).$$

Consequently $T_n(f) = g$, that is T_n is surjective.

It remains to prove the continuity of the operator $T_n^{-1} : Z \to Z$. We shall prove sequential continuity, that is to say for each sequence $\{T_n(f_j)\}_{j \in \mathbb{N}} \subset Z$ with $\{f_j\}_{j \in \mathbb{N}} \subset Z$ such that $\lim_{j \to \infty} T_n(f_j) = T_n(f)$, we have $\lim_{j \to \infty} T_n^{-1}(T_n(f_j)) = T_n^{-1}(T_n(f))$ (or equivalently $\lim_{j \to \infty} f_j = f$).

From the hypothesis $T_n(f_j) \to T_n(f)$ as $j \to \infty$ and from the definition of T_n, it follows that:

$$\|T_n(f_j) - T_n(f)\| = \max \Big\{ \sup_{x \in C_1} \|f_j(x) - f(x)\|, \qquad (*)$$

$$\max_{2 \leq i \leq k_n} \sup_{x \in C_i} \|f_j(x) - f(x) + F(f_j)(r_n(x)) - F(f)(r_n(x))\| \Big\} \to 0 \quad \text{as} \quad j \to \infty.$$

Consequently the sequence $\{f_j\}_{j \in N}$ converges uniformly to f on C_1. Let us prove the uniform convergence also on C_2. To this aim set:

$$\varepsilon_j = \max \left\{ \frac{1}{j}, \sup_{x \in C_1} \|f_j(x) - f(x)\| \right\}.$$

The definition of $\{\varepsilon_j\}_{j \in N}$ implies that $\lim_{j \to \infty} \varepsilon_j = 0$ and $\operatorname{Im}(f_j - f)|_{C_1} \subseteq B(0, \varepsilon_j) \subset Y$.
Consider the function $\phi_j : K \to B(0, \varepsilon_j)$ defined by the formula:

$$\phi_j(x) = (f_j - f)(\rho(x)), \qquad \text{with } x \in K,$$

$$\rho(x) \equiv \begin{cases} x & \text{for } x \in C_1 \\ \dfrac{x}{n\|x\|} & \text{for } x \in K \setminus C_1. \end{cases}$$

Since for $\|x\| = \frac{1}{n}$ we have $\rho(x) = \frac{x}{n \cdot \frac{1}{n}} = x$, the function ϕ_j becomes a continuous extension of the restriction $(f_j - f)|_{C_1}$. Observe that:

$$\|\rho(x)\| = \left\| \frac{x}{n\|x\|} \right\| = \frac{1}{n} \quad \forall\, x \in K \setminus C_1 \quad \text{so} \quad \rho(x) \in C_1 \quad \forall\, x \in K.$$

But on C_1 we have $f_j \to f$, which implies $\phi_j(x) \to 0$ as $j \to \infty$ on K.
From the continuity of F it follows that

$$\lim_{j \to \infty} F(f + \phi_j) = F(f) \quad \text{on} \quad K.$$

By the definition of ϕ_j we have $(f + \phi_j)|_{C_1} \equiv (f_j)|_{C_1}$ and Hypothesis ii) implies:

$$F(f + \phi_j)|_{C_1} \equiv F(f_j)|_{C_1}.$$

Consequently $\lim_{j \to \infty} F(f_j) = F(f)$ on C_1. From property P_3) we have $r_n(x) \subseteq C_1 \quad \forall x \in C_2$ and so

$$\|F(f_j)(r_n(x)) - F(f)(r_n(x))\| \to 0 \quad \text{as} \quad j \to \infty \quad \text{on} \quad C_1 \cup C_2.$$

Comparing this expression with $*$) we see that:

$$\lim_{j \to \infty} \|f_j - f\| = 0 \quad \text{on} \quad C_2.$$

Therefore we have proved that $f_j \to f$ uniformly on $C_1 \cup C_2$. A similar argument on $C_3, \ldots, C_{k_n}$ implies the uniform convergence of f_j on K and the continuity of T_n^{-1}.

q.e.d.

Now let us investigate the application of Theorems 4.3 and 4.4 to the Cauchy problem. To this end we shall replace the hypothesis of continuity of the function $f : I \times X \equiv [t_0, t_0 + a] \times X \to X$ with the classical Caratheodory conditions:

c$_1$) $\forall\, x \in X$ the function $f(\cdot, x)$ is strongly measurable

c$_2$) for a.e. $t \in I$ the function $f(t, \cdot)$ is continuous

c$_3$) $\exists$ an element $h \in L^1(I)$ such that $\forall\, x \in X$ we have $\|f(t,x)\| \leq h(t)$ with $t \in I$.

Under these conditions, a solution of the problem C) will be a function $x : J \equiv [t_0, t_0 + b] \subset I \to X$ satisfying the following conditions:

i) $x(t_0) = x_0$

ii) x is absolutely continuous on J

iii) x is differentiable on J, that is $\lim_{h \to 0} \frac{x(t+h) - x(t)}{h}$ exists a.e. in $int\, J \equiv (t_0, t_0 + b)$.

iv) $\dot{x}(t) = f(t, x(t))$ for a.e. $t \in J$.

Notice that in a Banach space the condition ii) does not imply iii); the two assumptions are equivalent only in reflexive or finite dimensional spaces.

Now we can prove the following result:

Theorem 4.5. *Let X be a Banach space, $x_0 \in X$ and $I \equiv [t_0, t_0 + a] \subset \mathbb{R}$. Suppose that the function $f : I \times X \to X$ satisfies the Caratheodory hypotheses and that there exists a subinterval $J \equiv [t_0, t_0 + b] \subset I$ such that the integral operator $F(x)(t) = x_0 + \int_{t_0}^{t} f(s, x(s))\,ds$ satisfies the Palais-Smale condition on $C(J, X)$.*

Then the problem C) admits a solution defined on at least J and the set of solutions on J is a compact R_δ.

Proof. From the properties of the Bochner integral it follows immediately that F is continuous on $C(J, X)$, even if in general it is not compact. Moreover F satisfies conditions i) and ii) of Theorem 4.4. In fact:

i) $F(x)(t_0) = x_0 + \int_{t_0}^{t_0} f(s, x(s))\,ds = x_0 \quad \forall\, x \in C(J, X)$

ii) If $g|_{K_\varepsilon} \equiv u|_{K_\varepsilon}$, then

$$F(g)(t) = x_0 + \int_{t_0}^{t} f(s, g(s))\,ds = x_0 + \int_{t_0}^{t} f(s, u(s))\,ds = F(u)(t) \quad \forall\, t \in K_\varepsilon.$$

As h is Lebesgue integrable on $\mathbb{R}^+$, the following relation holds:

$$\forall\,\varepsilon > 0 \; \exists\, \delta > 0 \quad \text{such that } \forall \text{ Borelian } B \subset J \text{ with } \mu(B) < \delta \text{ we have } \int_B h(t) < \varepsilon.$$

Then, by Hypothesis c_3), given $t,\,s \in J$ with $t \leq s$ and $t - s < \delta$, $\forall\, x \in C(J, X)$ we have:

$$\|F(x)(t) - F(x)(s)\| = \left\|\int_s^t f(z, x(z))\, dz\right\|$$

$$\leq \int_s^t \|f(z, x(z))\|\, dz \leq \int_s^t h(z)\, dz < \varepsilon.$$

This proves Hypothesis iii) of Theorem 4.4.

Since F is continuous and satisfies the Palais-Smale condition, the operator $T = I - F$ is 0-closed (see [50]). Therefore Theorem 4.3 states that the problem C) admits at least one solution thus the solution set $S \neq \phi$. From the remark to Theorem 4.1 we have:

$$\lim_{n \to \infty} \alpha(S_n) = 0 \quad \text{where} \quad S_n = \left\{x \in X : \|F(x) - x\| \leq \frac{1}{n}\right\}.$$

This relation, together with $S \neq \emptyset$, implies that S is totally bounded in $C(J, X)$ (see [50]). Finally, from the continuity of F it follows that S is closed and so compact. By applying Theorem 4.3, we see that the solution set is a compact R_δ.

q.e.d.

We remark that other mathematicians have obtained Aronszajn type theorems in the following cases:

a) $f : I \times B \to X$ with $B = B(x_0, r)$ a ball and f satisfying Caratheodory conditions,

b) $f : I \times B \to X$ with $B = B(x_0, r)$ where f is continuous and bounded.

In both cases a solution of C) is a function $x : J \subset \mathbb{R} \to B(x_0, r)$. Nevertheless, by defining a function $\rho : X \to B(x_0, r)$ as:

$$\rho(x) = \begin{cases} x & \text{if } x \in B \\ \dfrac{rx}{\|x\|} & \text{if } \|x\| \geq r \end{cases}$$

and setting $g(t, x) = f(t, \rho(x))$, we obtain a function $g : I \times X \to X$ which is a continuous extension of f and satisfies the Caratheodory conditions in both cases

a) and b). Naturally, to each solution of the problem C) relative to f corresponds one solution of the problem C) relative to g and vice versa (the intervals of existence may differ). Therefore the cases a) and b) can be reduced to the preceding theorem and there is not any loss of generality in supposing that f is defined everywhere on $I \times X$.

In conclusion, we present the other (few) existing results of Aronszajn type, almost all due to Szufla, and try to connect them to Theorem 4.5.

We begin with the situation a). Let $H : \mathbb{R} \to \mathbb{R}^+$ be a function defined as:

$$H(t) = \begin{cases} 0 & \text{if } t < t_0 \\ \int_{t_0}^{t} h(s)ds & \text{if } t \in I \equiv [t_0, t_0 + a] \\ \int_{t_0}^{t_0+a} h(s)ds & \text{if } t > t_0 + a. \end{cases}$$

Let c be a constant which is a solution of the equation $H(t_0 + c) = r$ if this solution exists, otherwise set $c = a$. Then the following result of Szufla (see [134]) holds:

Theorem 4.6. *Assume the hypotheses of case a). Let $b = \min\{a, c\}$, $J \equiv [t_0, t_0 + b] \subset \mathbb{R}$, $p \in (0, b]$, $J_p \equiv [t_0, t_0 + p]$ and $Q_n^p = \{x \in C(J_p, B) : x(t_0) = x_0$ and $\|x - F(x)\| \leq \frac{1}{n}\}$ where F is the integral operator relative to f. Suppose that $\lim_{n \to \infty} \alpha(Q_n^p) = 0 \quad \forall p \in (0, b]$.*
Then the solution set of C) is a compact R_δ in $C(J, X)$.

Sketch of Proof. The set $X_1 \equiv \{x \in C(J, X) : x(t_0) = x_0\}$ is closed in $C(J, X)$. Szufla, using the properties of the measure of noncompactness, succeeded in showing that the condition $\lim_{n \to \infty} \alpha(Q_n^p) = 0$ implies $\lim_{n \to \infty} \alpha(S_n) = 0$.

But this last is equivalent to the Palais-Smale condition by the remark to Theorem 4.1; therefore the conclusion follows from Theorem 4.5.

q.e.d.

We now consider a Theorem of Szufla (see [135]) from 1974:

Theorem 4.7. *Assume the hypotheses of case b). Let $M = \sup_{(t,x) \in I \times B} \|f(t, x)\|$, $b = \min\{a, \frac{r}{M}\}$, $J \equiv [t_0, t_0 + b]$, $J_p \equiv [t_0, t_0 + p]$ with $p \in (0, b]$. Suppose that $\lim_{n \to \infty} \alpha(S_n^p) = 0$ where $S_n^p = \{x \in C^1(J_p, X) : x(t_0) = x_0, \|\dot{x} - f(\cdot, x)\|_{J_p} \leq \frac{1}{n}$ and $x(J_p) \subseteq B\} \quad \forall p \in (0, b]$. Then the solution set is a compact R_δ in $C(J, X)$.*

Proof. As in the preceding case, Szufla first proves that $\lim_{n\to\infty} \alpha(S_n) = 0$ where $S_n \subset C^1(J,X)$ $\forall\, n \in N$. Thus Theorem 4.5 implies that S is a compact R_δ on $C^1(J,X)$.

Let us take a sequence $\{x_n\}_{n\in N} \subseteq S$ with $x_n \to x$ in the norm of $C(J,X)$. As usual, $F(x) \equiv x_0 + \int_{t_0}^{t} f(s, x(s))ds$. Then from the continuity of F and f we have:

$$\|x_n - x\|_1 = \|F(x_n) - F(x)\|_1$$
$$\equiv \|F(x_n) - F(x)\| + \|f(\cdot, x_n) - f(\cdot, x)\| \to 0 \quad \text{as} \quad n \to \infty$$

that is, $\lim_{n\to\infty} x_n = x$ in the norm of $C^1(J,X)$. It is clear that the condition $\|x_n - x\|_1 \to 0$ always implies $\|x_n - x\| \to 0$. Therefore the spaces $(S, \|\cdot\|)$ and $(S, \|\cdot\|_1)$ are homeomorphic, which means that S is a compact R_δ in $C(J,X)$.

q.e.d.

We note that there exists another result relative to case b) which differs from Theorem 4.7 only in small variations on the assumptions about the range of f (see Szufla [136]).

Observe that, since each compact R_δ is also a continuum, all of the quoted results imply that the set S is nonempty, compact and connected. In fact *every known existence theorem guarantees both the continuity and the R_δ-compactness of* S, as is explained in Szufla [142] and [143].

From this point of view the Palais-Smale hypothesis (or the equivalent $\lim_{n\to\infty} \alpha(S_n) = 0$) represents the most general existence condition.

Ricceri has extended some of the results of Szufla, but the hypotheses are too complex to describe here, we refer the reader to [116].

We conclude this chapter with a further clarification of the structure of Peano funnel. At first glance it seems that with some more effort one could prove that the solution set is not only a compact R_δ, but a contractible set. Actually this is not possible even in finite dimensional spaces, as we can see from the following counterexample in $\mathbb{R}^2$ taken from Horst [65]:

Example 4.1. Let $t^+ = \max\{t, 0\}$ and $f : \mathbb{R}^2 \to [0, +\infty)$ a continuous function defined as:

$$f(t,z) = \begin{cases} 0 & \text{for } t \le 0 \text{ or } z \le 0 \\ 2t & \text{for } t > 0 \text{ and } z > t^2 \\ \dfrac{2z}{t} & \text{for } t > 0 \text{ and } 0 < z \le t^2. \end{cases}$$

The function f is continuous on the points of the parabola $z = t^2$ and so everywhere in $\mathbb{R}^2$. We have $f(t,z) \le 2t^+$ $\forall\, (t,z) \in \mathbb{R}^2$. It is known that the Cauchy problem:

$$\begin{cases} \dot{z}(t) = f(t, z(t)) \\ z(0) = 0 \end{cases} \qquad *)$$

admits on $[0,1]$ solutions of the form $z_\alpha(t) = \alpha t^2$ with $\alpha \in [0,1]$. Let $\beta,\, \gamma \in I \equiv [0,1]$ and $(t,x) \in \mathbb{R}^2$. Set:

$$g(t, w, \beta, \gamma) = (1 - \beta) \cdot f(t, w) + 2\beta\gamma t^+.$$

Since g is linear combination of continuous functions, it will be continuous on $\mathbb{R}^2 \times I^2$. From the relation $\sup_{\beta,\gamma \in I} 1 + \beta(\gamma - 1) = 1$ we have:

$$0 \le g(t, w, \beta, \gamma) \le (1-\beta)2t^+ + 2\beta\gamma t^+ = [1 + \beta(\gamma - 1)]2t^+ \le 2t^+.$$

Consider the new problem:

$$\begin{cases} \dot{w} = g(t, w, \beta, \alpha) \\ w(0) = 0 \end{cases} \qquad **)$$

where $\beta \in (0,1]$ and $\alpha \in I$ are fixed. As $0 \le z_\alpha(t) \le t^2$ $\forall\, t \in I$, we have:

$$g(t, z_\alpha, \beta, \alpha) = \begin{cases} (1-\beta)\cdot 0 + 2\beta \cdot 0 & \text{for } t \le 0 \\ (1-\beta)\dfrac{2z_\alpha(t)}{t} + 2\alpha\beta t^+ = 2\alpha t(1-\beta) + 2\alpha\beta t = 2\alpha t & \text{for } t > 0. \end{cases}$$

Hence:

$$\dot{z}_\alpha(t) = 2\alpha t = g(t, z_\alpha, \beta, \alpha) \quad \text{on} \quad I$$

that is z_α is a solution of $**)$ on I. We want to show it is unique.

Suppose that there exists another solution $y(t)$. From $0 \leq g(t, y(t), \beta, \alpha) \leq 2t$ we obtain the inclusion:

$$\text{graph of } y(t) \equiv \{(t, y(t)) : t \in I\} \subseteq D = \{(t, x) \in I^2 : 0 \leq x \leq t^2\}.$$

Set $y_1 = y(1)$. Since $(1-\beta) \cdot f(t, y(t)) = 2(1-\beta) \frac{y(t)}{t}$, $y(t)$ must be a solution on $(0, 1]$ of the Cauchy problem:

$$\begin{cases} \dot{y}(t) = 2(1-\beta) \frac{y(t)}{t} + 2\alpha\beta t \\ y(1) = y_1. \end{cases}$$

This is a linear differential equation, so it admits an unique solution $u(t)$ of the form:

$$u(t) = (y_1 - \alpha)t^{2(1-\beta)} + \alpha t^2 \equiv y(t) \quad \text{with} \quad t \in (0, 1].$$

As $\beta > 0$, the graph of $y(t)$ remains in D only if $\alpha = z_1$. Therefore the function $\alpha t^2 = z_\alpha(t)$ is the unique solution of $**$) on I.

Finally consider the Cauchy problem in $\mathbb{R}^2$:

$$\begin{cases} \dot{x}(t) = f(t, x(t)) \\ \dot{y}(t) = \left(1 - \frac{x(t)}{\phi(t)}\right) f\left(t - \frac{1}{2}, y(t)\right) + 2\sin^2\left(\frac{t^2}{x(t)}\right) \max\left\{0, t - \frac{1}{2}\right\} \frac{x(t)}{\phi(t)} \quad ***) \\ x(0) = y(0) = 0 \end{cases}$$

where $\phi(t) = \max\{\frac{1}{4}, t^2\}$. Defining $0 \cdot \sin^2(\frac{t^2}{0}) = 0$ makes the right hand side continuous. Since every continuous function in the finite dimensional situation always satisfies Hypothesis a) of Theorem 3.4, the solution set of $(***)$ is connected.

Let $(x(t), y(t))$ be a solution of $(***)$, then we have $x(t) = z_\alpha(t)$ with $\alpha \in [0, 1]$. By definition of $\dot{y}(t)$ we obtain:

$$\dot{y}(t) = 0 \quad \text{for} \quad 0 \leq t \leq \tfrac{1}{2}, \quad \text{so} \quad y(t) \equiv 0 \quad \text{on} \quad [0, \tfrac{1}{2}].$$

If $\alpha = 0$ then $x(t) \equiv 0$ and

$$\dot{y}(t) = \left(1 - \frac{0}{\phi(t)}\right) f\left(t - \tfrac{1}{2}, y(t)\right) = f\left(t - \tfrac{1}{2}, y(t)\right).$$

This expression represents the differential equation of the problem $*)$ with variable $t - \frac{1}{2}$. Therefore there exists a $\delta \in I$ such that:

$$y(t) = z_\delta(t - \tfrac{1}{2}) \quad \text{for} \quad t \in [\tfrac{1}{2}, 1].$$

Let $\alpha \in (0,1]$. Then for $t \in [\frac{1}{2},1]$ we have $\phi(t) \equiv t^2$ and

$$\dot{y}(t) = (1-\alpha)f(t-\tfrac{1}{2}, y(t)) + 2\alpha(t-\tfrac{1}{2})\sin^2(\tfrac{1}{\alpha}).$$

This is in fact the differential equation of the problem **) translated by $\frac{1}{2}$. Consequently, from the uniqueness of solutions of **), it follows that

$$y(t) = z_{\sin^2(\frac{1}{\alpha})}(t-\tfrac{1}{2}) \quad \text{for } t \in [\tfrac{1}{2},1].$$

In conclusion, we have the following situation:

$x(t) = 0 = z_0(t)$ and $y(t) = z_\delta(t-\tfrac{1}{2})$ with $\delta \in [0,1]$ for $\alpha = 0$

$x(t) = z_\alpha(t)$ and $y(t) = z_{\sin^2(\frac{1}{\alpha})}(t-\tfrac{1}{2})$ for $\alpha \in (0,1]$.

Now it is obvious that the solution set of ***) is homeomorphic to:

$$\{(0,\delta) : \delta \in [0,1]\} \cup \{(\alpha, \sin^2(\tfrac{1}{\alpha})) : \alpha \in (0,1]\}.$$

This set represents the well-known classical example of a set that is connected, but not locally or arcwise connected. It follows that the Peano funnel cannot be a contractible set.

5. Differential Inclusions in Banach Spaces

Let X be a Banach space, $\mathcal{P}(X)$ the power set of X, $T \equiv [0, b] \subset \mathbb{R}$, $x_0 \in X$ and $F(t, x)$ a multifunction from $T \times X$ into $\mathcal{P}(X)$.

Consider the differential inclusion:

$$\begin{cases} \dot{x}(t) \in F(t, x(t)) \\ x(0) = x_0. \end{cases} \qquad \text{M)}$$

By a solution of the problem M) we mean an absolutely continuous function $x : T \to X$ satisfying M) almost everywhere on T.

Beginning in the seventies the multivalued Cauchy problem in abstract spaces has been studied by many authors: we mention the existence theorems obtained by Chow and Schuur [27], Muhsinov [93], De Blasi [33], Anichini and Zecca [3], Sentis [125], Pavel and Vrabie [109], Tolstonogov [147] and [148] and Kisielewicz [77].

As usual, the first articles on the topological structure of the solution set S of the problem M) deal with the finite dimensional case. First Davy [31] proved that the set S is a continuum in $C(T, \mathbb{R}^n)$. Later Lasry and Robert [88] showed that S is acyclic whenever F has compact and convex values and is Hausdorff-u.s.c. Subsequently Himmelberg and Van Vleck [64] proved that the set S is a compact R_δ when F is u.s.c. and bounded. De Blasi and Myjak [36] extended this result to differential inclusions in which F satisfies the Caratheodory conditions, that is $F(\cdot, x)$ is measurable
$\forall\, x \in B(x_0, r)$ and $F(t, \cdot)$ is u.s.c. almost everywhere on $[t_0, t_0 + a]$.

We mention also the finite dimensional results obtained by Plaskacz [114], Papageorgiou [108], Gorniewicz [60] and Bogatyrev [14]. The first showed that the solution set is an R_δ when $x_0 \in C \subset \mathbb{R}^n$ with C a closed set and x_0 is not necessarily an interior point of C; the second treated a differential inclusion with time varying state constraints and the last one proved, in very particular cases, that the solution set of M) is a contractible set.

Only recently has the analogous problem been studied in abstract spaces. We note that in these studies the method of proof usually tries to follow the technique for the result in finite dimensions, adding hypotheses which guarantee the existence of solutions of M). In this way in 1982 Tolstonogov (see [149] and [150]) generalized a theorem of Davy, and later Papageorgiou [106] extended to separable Banach spaces a result of Himmelberg and Van Vleck. We shall present an infinite dimensional version of De Blasi and Myjak's theorem. Gel'man [53] has proved results on the connectedness, compactness, acyclicity and non-emptiness of the set of fixed points of multivalued operators between topological spaces.

By $\mathcal{P}_K(X)$ and $\mathcal{P}_{KC}(X)$ we shall mean the metric spaces composed respectively of the compact subsets of X, and the compact and convex subsets of X. For a metric we use the Hausdorff metric (note $h(A, \emptyset) = +\infty$):

$$h(A, C) = \inf\{r > 0 : A \subseteq B(C, r) \text{ and } C \subseteq B(A, r)\}.$$

The following continuous selection property holds for multifunctions (see [77]):

Theorem 5.1. *Let X be a separable Banach space, $T \equiv [0, b] \subset \mathbb{R}$ and $\mathcal{B}(T, X)$ the space of Bochner integrable functions from T to X. Suppose that $F : T \times X \to \mathcal{P}_{KC}(X)$ satisfies the Caratheodory conditions: that is, $F(\cdot, x)$ is a measurable map and $F(t, \cdot)$ is u.s.c.*

Then there exists a continuous function $f : X \to \mathcal{B}(T, X)$ such that $f(x)(t) \in F(t, x) \quad \forall\, x \in X$ and almost everywhere on T.

Using Theorem 5.1 and a result quoted in Chapter 2, we can considerably simplify the proof of an existence theorem due to Kisielewicz:

Theorem 5.2. *Let X be a separable Banach space, $x_0 \in X$, $T \equiv [0, b] \subset \mathbb{R}$ and $F : T \times X \to \mathcal{P}_{KC}(X)$ a multifunction satisfying the Caratheodory conditions. Assume that for each bounded set $A \subset X$ we have $\chi(F(t, A)) \leq g(t, \chi(A))$ almost everywhere on T where $g : T \times \mathbb{R}^+ \to \mathbb{R}^+$ is a Kamke function, and χ is the Hausdorff measure of noncompactness.*

Then the problem M) admits at least one solution.

Proof. By Theorem 5.1 there exists a continuous function $f : X \to \mathcal{B}(T, X)$ such that $f(x)(t) \in F(t, x) \quad \forall\, x \in X$ almost everywhere on T. Moreover:

$$\chi(f(A)(t)) \leq \chi(F(t, A)) \leq g(t, \chi(A)) \quad \forall \text{ bounded } A \subset X \qquad *)$$

Now consider the initial value problem:

$$\begin{cases} \dot{x}(t) = \tilde{f}(t, x(t)) \\ x(t) = x_0 \end{cases} \quad C^*)$$

where $\tilde{f}(t, x(t)) = f(x)(t)$.

As X is separable, the condition $*)$ is sufficient to guarantee the existence of an absolutely continuous function $x : T \to X$ which is a solution of $C^*)$ (see Song [127]). Then we have:

$$\dot{x}(t) = \tilde{f}(t, x(t)) = f(x)(t) \in F(t, x(t)) \quad \text{and} \quad x(0) = x_0$$

that is, x is a solution of M).

q.e.d.

Remark. If F is continuous in x then it is sufficient in Theorems 5.1 and 5.2 to require that $F : T \times X \to \mathcal{P}_K(X)$.

The next theorem, taken from De Blasi [35], is useful for the approximation of multifunctions in abstract spaces:

Theorem 5.3. *Let $F : X \to \mathcal{P}_{KC}(X)$ be a multifunction. Then the following conditions are equivalent:*
a) *F is compact h-u.s.c*
b) *there exists a sequence $\{G_n\}_{n \in N}$ of compact continuous multifunctions $G_n : X \to \mathcal{P}_{KC}(X)$ satisfying the properties:*
   b_1) $G_1(x) \supset G_2(x) \supset \ldots$
   b_2) $F(x) \subset G_n(x) \quad \forall n \in N$
   b_3) $G_n(x) \subset \overline{co} \bigcup \{F(x) : x \in X\} \forall n \in N$
   b_4) $h(G_n(x), F(x)) \to 0$ as $n \to \infty$.

As a preliminary to our next theorem, we state and sketch the proof of a result due to Rzezuchowski [121], which in turn improves on a result of Jarnik and Kurzweil [70]. The theorem is in part a multifunction version of a well-known theorem of Scorza-Dragoni, and is in a certain sense an analogue of Lusin's famous theorem that every measurable function is "almost" continuous.

Theorem 5.4. *Let $F: T \times X \to \mathcal{P}_C(Y)$, with T a measurable subset of $\mathbb{R}$, X and Y separable metric spaces. Assume that for almost all $t \in T$, the graph of $F(t, \cdot)$ is closed in $X \times Y$. Then there exists a multifunction with closed values*

$$F^*: T \times X \to \mathcal{P}(Y)$$

such that
  (1) *for almost all $t \in T$, $F^*(t, x) \subset F(t, x)$ for all $x \in X$;*
  (2) *if $\Delta \subset T$ is measurable and $u: \Delta \to X$, $v: \Delta \to Y$ are measurable functions with $v(t) \in F(t, u(t))$ a.e. in Δ, then $v(t) \in F^*(t, u(t))$ a.e. in Δ;*
  (3) *$\forall \varepsilon > 0 \ \exists$ a closed $T_\varepsilon \subset T$ with $\mu(T \setminus T_\varepsilon) < \varepsilon$ and the graph of $F^*|_{T_\varepsilon \times X}$ is closed in $T \times X \times Y$.*

Remark. The connection with u.s.c. is based on the fact that if a multifunction has closed values and closed domain, then u.s.c. $\Longrightarrow$ closed graph. The converse is true if the closure of the range of $F \ (= \bigcup_{x \in X} F(x))$ is a compact subset of Y.

Sketch of Proof. First recall Theorem IV.8.22 from Dunford and Schwartz [51] which asserts that "every function is almost measurable";

Let $\varphi: T \to \mathbb{R}_+ \cup \{\infty\}$ be an arbitrary function.

Then there exists a measurable function $\varphi^*: T \to \mathbb{R}_+ \cup \{\infty\}$ such that

$$\varphi \leq \varphi^*(t) \quad \text{a.e. in } T; \tag{1}$$

$$\text{if } \varphi(t) \leq \psi(t) \quad \text{a.e. in } T \text{ with } \psi \text{ measurable,} \tag{2}$$
$$\text{then} \quad \varphi^*(t) \leq \psi(t) \quad \text{a.e. in } T;$$

Now let $\Phi(t) = Gr(F(t, \cdot)) \equiv \bigcup_{x \in X} \bigcup_{y \in F(t,x)} (x, y)$. Note that $(x, y) \in \Phi(t)$ if and only if $y \in F(t, x)$. Let $\{a_i\} \subset X \times Y$ be a countable dense subset and define $\varphi_i(t) = \tilde{h}(a_i, \Phi(t))$ ($\tilde{h}(a_i, \emptyset) = +\infty$). By the result quoted above, there exists a measurable φ_i^* with properties (1) and (2). Define $\Phi^*(t) = \bigcap_{i=1}^{\infty} \{(x, y) \mid d(a_i, (x, y)) \geq \varphi_i^*(t)\}$. Then $F^*(t, x)$ is defined by

$$y \in F^*(t, x) \iff (x, y) \in \Phi^*(t).$$

Now we can prove the following result due to Papageorgiou [106]:

Theorem 5.5. Let X be a separable Banach space, $x_0 \in X$ and $T \equiv [0,b] \subset \mathbb{R}$. Suppose that the multifunction $F : T \times X \to \mathcal{P}_{KC}(X)$ satisfies the following conditions:

i) $F(\cdot, x)$ is measurable $\forall\, x \in X$
ii) $F(t, \cdot)$ is compact u.s.c. $\forall\, t \in T$ (i.e., if A is bounded, then $F(t, A)$ is compact, and F is u.s.c. on X)
iii) F maps compact sets of $T \times X$ to compact sets of X.
iv) $\|F(t, x)\| = \sup\limits_{y \in F(t,x)} \|y\| \leq a(t) + c(t)\|x\|$ almost everywhere on T with $a(\cdot)$ and $c(\cdot) \in L^1(\mathbb{R}^+)$.

Then the solution set S of the problem M) is a compact R_δ in $C(T, X)$.

Proof. First we look for a bound on the solutions of M). Let $x \in S$. Then we have $\dot{x}(t) \in F(t, x(t))$ and:

$$\|\dot{x}(t)\| \leq \|F(t, x(t))\| \leq a(t) + c(t)\|x\|$$

almost everywhere on T, so

$$\|x(t)\| \leq \|x_0\| + \int_0^t \|\dot{x}(t)\| dt \leq \|x_0\| + \int_0^t a(s) ds + \int_0^t c(s) \|x(s)\| ds$$

$$\leq \|x_0\| + b\|a\| + \int_0^t c(s) \|x(s)\| ds.$$

From Gronwall's Lemma it follows that

$$\|x(t)\| \leq (\|x_0\| + b\|a\|) \exp\left(\int_0^t c(s) ds\right)$$

$$\leq (\|x_0\| + b\|a\|) \exp(b\|c\|) = K.$$

Define the truncated mapping $\widetilde{F} : T \times X \to \mathcal{P}_{KC}(X)$ as:

$$\widetilde{F}(t, x) = \begin{cases} F(t, x) & \text{if } \|x\| \leq K \\ F(t, \frac{Kx}{\|x\|}) & \text{if } \|x\| > K. \end{cases}$$

Since $\widetilde{F}(t, x) = F(t, r(x))$ where $r : X \to B(0, K) \subset X$ is the continuous retraction of X such that $r|_{B(0,M)} \equiv \mathrm{Id}_X$, $\widetilde{F}$ has the same measurability and continuity properties as F.

The following inequalities hold:

$$\|\widetilde{F}(t,x)\| = \|F(t,x)\| \leq a(t) + c(t)\|x\| \leq a(t) + Kc(t) \qquad \forall \|x\| \leq K$$

$$\|\widetilde{F}(t,x)\| = \left\|F\left(t, \frac{Kx}{\|x\|}\right)\right\| \leq a(t) + c(t)\left\|\frac{Kx}{\|x\|}\right\| = a(t) + Kc(t) \quad \forall \|x\| > K.$$

Consequently $\|\widetilde{F}(t,x)\| \leq a(t) + Kc(t) \equiv \psi(t)$ with $\psi \in L^1(\mathbb{R}^+)$.

Consider the differential inclusion:

$$\begin{cases} \dot{x}(t) \in \widetilde{F}(t, x(t)) \\ x(0) = x_0. \end{cases} \qquad \widetilde{M})$$

Let $\widetilde{S}$ be the solution set of $\widetilde{M}$). If $x \in S$ then $\|x\| \leq K$; but F and $\widetilde{F}$ coincide on $B(0, K)$, so that $x \in \widetilde{S}$. Conversely, if $x \in \widetilde{S}$ we have:

$$\|\dot{x}(t)\| \leq a(t) + c(t)\|x(t)\| \quad \text{almost everywhere on} \quad T.$$

By again using Gronwall's inequality, we obtain:

$$\|x(t)\| \leq K \quad \text{and} \quad \dot{x}(t) \in F(t, x(t)), \quad \text{that is} \quad x \in S \quad \text{and} \quad S = \widetilde{S}.$$

Consequently, we can replace assumption iv) with the condition $\|F(t, x(t))\| \leq \psi(t)$, for all solutions $x(t)$, without loss of generality.

Using Theorem 5.4, we can assert that there exists $F^*(t, x)$ with properties (1), (2), (3) of that theorem. Define

$$\widehat{F}(t, x) = \overline{\text{co}}\, F^*(t, x).$$

Since $F^*(t, x) \subset F(t, x)$ and $F(t, x)$ is convex and compact, it follows that $\widehat{F}(t, x) \subset F(t, x)$; clearly $\widehat{F}$ has convex compact values.

Moreover, $\forall$ bounded $A \subset X$ we have $\widehat{F}(t, A) \subseteq F(t, A)$, so that $\widehat{F}$ is a compact mapping. Since Hypothesis ii) implies $\chi(F(t, A)) = 0$ for bounded subsets of X, we can apply to F Theorem 5.2, ensuring the existence of solutions to the problem M). Then conclusion (2) of Rzezuchowski's theorem implies that a.e. in t,

$$\dot{x}(t) \in \widehat{F}(t, x(t)).$$

Therefore $\widehat{F}(t, x(t)) \neq \emptyset$ almost everywhere on T and $x(0) = x_0$. If $\widehat{S}$ is the solution set of the problem M) relative to $\widehat{F}$, then the relation a) shows that $S = \widehat{S}$.

By property 3) of Rzezuchowski's theorem, for any $\varepsilon > 0$ there exists a closed $T_\varepsilon \subset T$ with $\mu(T \setminus T_\varepsilon) < \varepsilon$ such that $Gr(\widehat{F}|_{T_\varepsilon \times X})$ is closed in $T_\varepsilon \times X \times Y$. We want to show that for any sequences $\{(t_k, x_k)\}_{k \in N}$, $\{y_k\}_{k \in N}$, with $y_k \in \widehat{F}(t_k, x_k)$, $(t_k, x_k) \in T_\varepsilon \times X$,

$$\text{if } (t_k, x_k) \to (t, x) \text{ then } y_k \in \widehat{F}(t, x) + \varepsilon B(0, 1)$$

for any given $\varepsilon > 0$, $k > k_\varepsilon$. This will imply $\widehat{F}$ is Hausdorff u.s.c. on $T_\varepsilon \times X$, and because $\widehat{F}$ has compact values we can conclude that

$$\widehat{F} \text{ is u.s.c. on } T_\varepsilon \times X. \tag{*}$$

Suppose in fact that there exists a number $\varepsilon > 0$ and two sequences $\{(t_k, x_k)\}_{k \in N} \subset T_\varepsilon \times X$ and $\{y_k\}_{k \in N} \subset X$ with $\lim_{k \to \infty}(t_k, x_k) = (t, x)$ and $y_k \in \widehat{F}(t_k, x_k)$ such that $y_k \notin \widehat{F}(t, x) + \varepsilon B(0, 1) \quad \forall k \in N$. This is equivalent to $h(\widehat{F}(t_k, x_k), \widehat{F}(t, x)) \not\to 0$ as $k \to \infty$.

Then $\{y_k\}_{k \in N} \subseteq \bigcup_{k=1}^{\infty} \widehat{F}(t_k, x_k) \subseteq \bigcup_{k=1}^{\infty} F(t_k, x_k)$. Observe that the sequence $\{(t_k, x_k)\}_{k \in N}$ converges in $T_\varepsilon \times X$ and so it is a compact subset in $T \times X$. Therefore Hypothesis iii) implies the compactness of the set $\bigcup_{k=1}^{\infty} F(t_k, x_k)$. By passing to a subsequence if necessary, we can assume $y_k \to y$ as $k \to \infty$. But $Gr(\widehat{F}|_{T_\varepsilon \times X})$ is closed in $T_\varepsilon \times X \times Y$, so the cluster point (t, x, y) must belong, i.e., $y \in \widehat{F}(t, x)$, a contradiction.

So $h(\widehat{F}(t_k, x_k), \widehat{F}(t, x)) \to 0$ and $\widehat{F}$ is u.s.c. on $T_\varepsilon \times X$.

Now consider the following multifunction $\overline{F} : T \times X \to \mathcal{P}_{KC}(X)$:

$$\overline{F}(t, x) = c_{T \setminus T_0}(t) \widehat{F}(t, x) + c_{T_0}(t) F(t, x).$$

where $c_A(t)$ is the characteristic function of the set A and $T_0 = \{t \mid \widehat{F}(t, x) = \emptyset\}$. Similarly to our arguments involving $\widehat{F}$, recalling that F is u.s.c. everywhere on $T \times X$, it follows that

$$\forall \varepsilon > 0 \ \exists \ \overline{T}_\varepsilon \subseteq T \text{ with } \mu(T \setminus \overline{T}_\varepsilon) < \varepsilon \text{ such that } \overline{F}|_{\overline{T}_\varepsilon \times X} \text{ is u.s.c.}$$

Moreover, $\overline{F}$ and $\widehat{F}$ coincide on T, except for the points at which the equality $\widehat{F}(t, x) = \emptyset$ holds; clearly for such (t, x) we cannot have $\dot{x}(t) \in \widehat{F}(t, x)$ and thus (t, x) cannot belong to $\widehat{S} = S$. Therefore, if $\overline{S}$ denotes the solution set of M) relative to $\overline{F}$, we have $\widehat{S} \subseteq \overline{S}$. But $\overline{F}(t, x) \subseteq F(t, x)$, so $\overline{S} \subseteq S = \widehat{S}$ thus $\overline{S} = S$.

Let $T'_m = \{t \in T : m-1 \leq \psi(t) < m\}$ where $\psi \in L^1(\mathbb{R}^+)$ is the upper bound on $F(t, x(t))$. Then $T = \bigcup_{m=1}^{\infty} T'_m$.

Define the multifunction $F_m : T \times X \to \mathcal{P}_{KC}(X)$ as:

$$F_m(t, x) = c_{T'_m}(t) F(t, x).$$

Naturally we have $F(t, x) = \sum_{m=1}^{\infty} F_m(t, x)$ and $\|F_m(t, x)\| \leq m$. We shall call $\widehat{F}_m(t, x)$ the mapping related to $F_m(t, x)$ in the same way as $\widehat{F}(t, x)$ relates to $F(t, x)$. Again we conclude $\widehat{F}(t, x) = \sum_{m=1}^{\infty} \widehat{F}_m(t, x)$. Finally, consider the multifunction $\overline{F}_m(t, x) = c_{T \setminus T_0}(t) \widehat{F}_m(t, x) + c_{T_0}(t) F_m(t, x)$ where $\overline{F}(t, x) = \sum_{m=1}^{\infty} \overline{F}_m(t, x)$. By the arguments preceding (∗) above we know that there exists a set $T_m \subseteq T$ with $\mu(T \setminus T_m) \leq \frac{1}{2^m}$ such that $\overline{F}_m|_{T_m \times X}$ is u.s.c. Moreover, since F_m and $\widehat{F}_m$ are compact, $\overline{F}_m$ also is compact. Recalling that a u.s.c. mapping is always h-u.s.c. (see Aubin and Cellina [7]), Theorem 5.3 applied to each multifunction $\overline{F}_m$ implies the existence of compact mappings $G_n^m : T_m \times X \to \mathcal{P}_{KC}(X)$ such that:

$$\overline{F}_m(t, x) \subseteq G_n^m(t, x) \quad \forall n \in N,$$

$$G_1^m(t, x) \supseteq G_2^m(t, x) \supseteq \ldots \quad \text{and} \quad G_n^m(t, x) \xrightarrow{h} \overline{F}_m(t, x) \quad \text{as} \quad n \to \infty.$$

Set $W = T \setminus \bigcup_{m=1}^{\infty} T_m$, $A_1 \equiv T_1$ and $A_m = T_m \setminus \bigcup_{k=1}^{m-1} T_k$.

Since the sequence $\{T \setminus \bigcup_{k=1}^{m} T_k\}_{m \in N}$ is decreasing, we have $\mu(W) = \mu(T \setminus \bigcup_{m=1}^{\infty} T_m) = \mu(\bigcap_{m=1}^{\infty}(T \setminus T_m)) = \mu(\bigcap_{m=1}^{\infty}(T \setminus \bigcup_{k=1}^{m} T_k)) = \lim_{k \to \infty} \mu(T \setminus \bigcup_{k=1}^{m} T_k) \leq \lim_{m \to \infty} \frac{1}{2^m} = 0$, that is $\mu(W) = 0$. If $m \neq n$ (for example $m > n$) we have:

$$A_m \cap A_n = (T_m \setminus \bigcup_{k=1}^{m-1} T_k) \cap (T_n \setminus \bigcup_{k=1}^{n-1} T_k) \subseteq (T_m \setminus \bigcup_{k=1}^{m-1} T_k) \cap T_n = \emptyset$$

that is the sets A_m are mutually disjoint. Denoting $A = \bigcup_{m=1}^{\infty} A_m$, we see that:

$$T = (T \setminus \bigcup_{m=1}^{\infty} T_m) \cup (\bigcup_{m=1}^{\infty} T_m) = (T \setminus \bigcup_{m=1}^{\infty} T_m) \cup (\bigcup_{m=1}^{\infty} T_m \setminus \bigcup_{k=1}^{m-1} T_k) = W \cup A.$$

Now we introduce the following multifunction $G_n : T \times X \to \mathcal{P}_{KC}(X)$:

$$G_n(t,x) = c_N(t)\overline{F}(t,x) + \sum_{m=1}^{\infty} c_{A_m}(t) G_n^m(t,x).$$

From the definition of G_n it follows that $G_n(\cdot, x)$ is measurable, $G_n(t, \cdot)$ is compact and continuous (since the mappings G_n^m and $\overline{F}$ are) and $G_1(t,x) \supseteq G_2(t,x) \supseteq \cdots$.
Moreover,

$$\lim_{n\to\infty} G_n(t,x) = c_N(t)\overline{F}(t,x) + \sum_{m=1}^{\infty} c_{A_m}(t) \overline{F}_m$$

$$= c_N(t)\overline{F}(t,x) + c_A(t)\overline{F}(t,x) = \overline{F}(t,x).$$

Since the limit is relative to the Hausdorff metric and the sequence $\{G_n(t,x)\}_{n\in N}$ is decreasing, the equality $\overline{F}(t,x) = \bigcap_{n=1}^{\infty} G_n(t,x)$ holds.

Theorem 2 in Rybinski [117] states that we can find a selection $g_n : T \times X \to X$ of $G_n(t,x)$ which satisfies the Caratheodory conditions.

Notice that $\forall\, t \in T \setminus W = A$ we have $t \in A_m$ for some $m \geq 1$. Then $c_W(t) = 0$ and $c_{A_k}(t) = 0$ $\forall\, k \neq m$, so $g_n(t,x) \in G_n(t,x) \equiv G_n^m(t,x)$ $\forall\, x \in X$. In his proof of our Theorem 5.3 DeBlasi obtains the following expression for the multifunctions G_n^m :

$$G_n^m(t,x) = \sum_{k \in I_n^m} p_{nk}^m(t,x)\, G_{nk}^m$$

where I_n^m is a finite index set, $\{p_{nk}^m\}_{k \in I_n^m}$ is a locally Lipschitzian partition of unity and $G_{nk}^m \in \mathcal{P}_{KC}(X)$. Hence:

$$g_n(t,x) = \sum_{k \in I_n^m} p_{nk}^m(t,x)\, u_{nk}^m$$

where u_{nk}^m are points of G_{nk}^m. It follows that the function $g_n(t, \cdot)$ is locally Lipschitzian on $T \setminus W$. Now consider the differential inclusion:

$$\begin{cases} \dot{x}_n(t) \in G_n(t, x_n(t)) \\ x_n(0) = x_0. \end{cases} \qquad M_n)$$

Let S_n denote the solution set of the problem M_n), Theorem 5.2 implies $S_n \neq \emptyset$. Fix $x \in S_n$. $\forall\, r \in [0,1)$ let $\bar{x}_n^r$ be the unique solution of the following Cauchy problem defined on $[rb, b]$:

$$\begin{cases} \dot{\bar{x}}_n^r(t) = g_n(t, \bar{x}_n^r(t)) \\ \bar{x}_n^r(rb) = x(rb). \end{cases} \qquad \overline{C})$$

Then the function $y_n^r : T \to X$ defined as:

$$y_n^r(t) = \begin{cases} x(t) & \text{for } t \in [0, rb] \\ \bar{x}_n^r(t) & \text{for } t \in [rb, b] \end{cases}$$

is a solution $\forall\, r \in [0,1)$ of the problem M_n). In fact $y_n^r \equiv x \in S_n$ on $[0, rb]$, $\dot{y}_n^r = \dot{\bar{x}}_n^r = g(t, \bar{x}_n^r) \in G_n(t, \bar{x}_n^r)$ on $[rb, b]$ and $\lim_{t \to rb \pm} y_n^r(t) = x(rb)$. Finally consider the following function $h : [0,1] \times S_n \subset [0,1] \times C(T, X) \to C(T, X)$:

$$h_n(r, x) = \begin{cases} y_n^r & \text{for } r \in [0, 1), \\ x & \text{for } r = 1. \end{cases}$$

Here the function y_n^r is determined by the choice of $x \in S_n$. By definition we have $h_n(0, x) = \tilde{x}_n$ and $h_n(1, x) = x$. Since under our assumptions the solutions of abstract differential equations depend continuously on initial data (see Lakshmikantham and Ladas [85]), the function h is continuous on $(0,1) \times S_n$. Hence

$$\lim_{r \to 0+} h_n(r, x) = \bar{x}_n^0 \quad \forall\, t \in [0, b]$$

with $\bar{x}_n^0(0) = x(0) = x_0$ and $\dot{\bar{x}}_n^0(t) = g_n(t, \bar{x}_n^0(t))$.

The uniqueness of solutions of $\overline{C}$) determines $\bar{x}_n^0 \equiv \tilde{x}_n$. On the other hand, if $x \in S_n$, we see that:

$$\lim_{r \to 1-} h_n(r, x) = y_n^1 = x.$$

Therefore the mapping h_n is continuous on $[0,1] \times S_n$, so h_n is a homotopy and S_n is a contractible set $\forall\, n \in N$. Proceeding exactly as in the proof of Theorem 3.1 in [106], we see that S_n is also compact in $C(T, X)$.

It now remains to prove that $S = \bigcap_{n=1}^{\infty} S_n$.

Since $\overline{F}(t,x) \subseteq G_n(t,x) \ \forall \, n \in N$, we have $S \equiv \overline{S} \subseteq \bigcap_{n=1}^{\infty} S_n$. Conversely, let $x \in \bigcap_{n=1}^{\infty} S_n$, that is $\dot{x}(t) \in G_n(t,x(t)) \ \forall \, n \in N$ almost everywhere on T. Then:

$$\dot{x}(t) \in \bigcap_{n=1}^{\infty} G_n(t,x(t)) \equiv \overline{F}(t,x(t)) \quad \text{a.e., that is} \quad x \in \overline{S} \equiv S.$$

From Hyman's theorem it follows immediately that S is a compact R_δ.

q.e.d.

Notice that, if X is finite dimensional, each u.s.c. mapping with compact values is always compact u.s.c. Therefore Theorem 5.5 contains De Blasi and Myjak's result as a particular case. The assumption of compactness becomes indispensable in infinite dimensions in order to ensure the existence of solutions of the problem M).

Conti, Obukhovskii and Zecca [29] have investigated the topological structure of the set of mild solutions of the Cauchy problem $x'(t) \in A(t,x(t)) + F(t,x_t)$, $x(t) = x_0(t)$ for $t \in (\tau, 0]$ in a separable Banach space E. Here $x_t(\cdot) = \{x(t+s) | -\tau \leq s \leq 0\}$, $x_0(\cdot)$ is given and $\{A(t)\}_{t \in [0,T]}$ is a family of closed linear (not necessarily bounded) operators in E generating a strongly continuous evolution operator $K : \Delta \to \mathcal{L}(E)$, where $\Delta = \{(t,s) | 0 \leq s \leq t \leq T\}$, which is continuous with respect to the norm of the space $\mathcal{L}(E)$ when $s < t$. They assume that the multifunction $F : [0,T] \times C([-\tau,0], E) \multimap \mathcal{P}_{KC}(E)$ satisfies:

A1) $\forall \, \phi \in C([-\tau,0], E)$, $F(\cdot, \phi) : [0,T] \to \mathcal{P}_{KC}(E)$ has a measurable selection;
A2) for a.e. $t \in [0,T]$, $F(t,\cdot) : C([-\tau,0), E) \multimap \mathcal{P}_{KC}(E)$ is u.s.c.;
A3) $\exists \, \alpha, \beta \in L^1([0,T])$, nonnegative, such that

$$\|F(t,\phi)\| \leq \alpha(t) + \beta(t) \|\phi\| \quad \forall \, \phi \in C([-\tau,0], E)$$

and a.e. $t \in [0,T]$;
A4) $\forall \, D \subset C([-\tau,0], E)$ which is nonempty, bounded and equicontinuous we have

$$\chi(F(t,D)) \leq k(t) \sup_{[-\tau,0]} \chi(D(t))$$

where $D(t) = \{\phi(t) | \phi \in D\}$, and $k \in L^1([0,T])$, and χ is the Hausdorff measure of noncompactness in $C([-\tau,0], E)$.

A function $x(\cdot) \in C([-\tau, T], E)$ is a mild solution of the above Cauchy problem if $x(t) = x_0(t)$ for $t \in [-\tau, 0]$ and $x(t) = K(t,0)x(0) + \int_0^t K(t,s)f(s)ds$ for $t \in [0,T]$ where $f(\cdot)$ is a Bochner integrable selector of $F(t, x_t)$. Obukhovskii [103] proved that under these assumptions the set of all mild solutions of the above Cauchy problem is nonempty and compact in $C([-\tau, T], E)$.

Theorem 5.6. (Conti, Obukhovskii and Zecca [29]) *Under the above assumptions, the set of mild solutions is an R_δ-subset of $C([-\tau, T], E)$.*

As in the case of ordinary differential equations, questions about the solution set of M) are questions about the set Fix $(T) = \{x | x \in T(x)\}$ of fixed points of a multifunction $T : X \multimap X$, X a linear topological space. Gel'man [53], Ricceri [115], Bressan, Cellina and Fryszkowski [16], and Kánnai and Tallos [72] have provided useful tools.

Gel'man [53] discusses the structure of the set of fixed points for a multivalued operator T on a Banach space X. Results of this type can be applied to M) by creating the sets

$$\mathcal{K} = \{x \in C(I; X) \mid x(\cdot) \text{ is Lipschitz}, \; x(t_0) = x_0\},$$
$$c\ell(x) = \{z \in \mathcal{K} \mid z'(t) \text{ exists and } z'(t) \in F(t,x) \text{ a.e. on } I\}.$$

Then a fixed point of the map $T : x \mapsto c\ell(x)$ is a solution of M). In order to describe his results, we recall (see [53]) that one can define an integer-valued index for a multivalued vector field of the form $\varphi = Id - T$, where Id is the identity map and $T : X \multimap X$ is an upper semicontinuous multifunction with closed values and $\overline{T}(X)$ compact. This number, denoted $\gamma(\varphi, \partial U)$, will exist for any open bounded set U as long as $0 \notin \varphi(y) \; \forall \, y \in \partial U$, and will have all the standard properties of an index. In particular, $\gamma(\varphi, \partial U) \neq 0$ implies $\exists \, x_0 \in U$ such that $x_0 \in T(x_0)$, i.e., we have a fixed point for T.

Theorem 5.7. (Gel'man [53]) *Let X be a Banach space, $U \subset X$ open and bounded, $T : \overline{U} \to \mathcal{P}_{KC}(X)$ upper semicontinuous with $\overline{T}(U)$ compact. Assume T is Hausdorff-metric Lipschitz on $\overline{U}$ with Lipschitz constant 1, and that $\gamma(Id - T, \partial U) \neq 0$. Then the set of fixed points of T, Fix (T), is nonempty and connected. If in addition $\dim T(x) \geq n \; \forall \, x \in \overline{U}$, then Fix (T) is also compact and $\dim$ Fix $(T) \geq n$.*

Theorem 5.8. (Ricceri [115]) *If $T : X \multimap X$ is a multifunction on a Banach space X, with closed convex bounded values, and if T is a contraction with respect to the Hausdorff metric, then $\operatorname{Fix}(T)$ is an absolute retract (hence arcwise connected).*

We recall that a subset A of $L^1(J, X)$ is *decomposable* if $\forall\, u, v \in A$, $\forall$ measurable $U \subset J$, $u\chi_U + v\chi_{J\setminus U} \in A$.

Theorem 5.9. (Bressan, Cellina and Fryszkowski [16]) *Assume that $X = L^1(J, E)$ where E is a Banach space and J is a measure space with finite positive nonatomic measure μ. Assume that X is separable, and T is a contraction, with nonempty, bounded, closed decomposable values. Then $\operatorname{Fix}(T)$ is an absolute retract.*

We note that a result similar to Theorem 5.5 has been proved by Deimling (see [45]) using a different approximation theorem taken from his monograph on functional analysis ([44]).

Lim [89] has proved a useful result on the stability of fixed points of multivalued contractions:

Theorem 5.10. (Lim [89]) *Let (X, d) be a complete metric space, and assume we are given maps $T_i \colon X \multimap X$, $i = 1, 2$ with nonempty closed values, satisfying*

$$h(T_i(x), T_i(y)) \leq \gamma d(x, y) \quad \forall\, x, y \in X,$$

for some $\gamma \in (0, 1)$. Then

$$h(Fix(T_1), Fix(T_2)) \leq \frac{1}{1 - \gamma} \sup_{x \in X} h(T_1(x), T_2(x)).$$

Kánnai and Tallos [72], improving on a result of Constantin [28], used Lim's theorem to prove the following.

Theorem 5.11. *For the problem M) on a separable metric space X assume*
(a) $F(t, x)$ *is measurable in t for each $x \in X$;*
(b) *there exists a locally integrable function $\ell(t)$ such that for a.e. $t \in [0, \infty)$ and all x, y in X,*

$$h(F(t, x), F(t, y)) \leq \ell(t) \|x - y\|,$$

$$\tilde{h}(0, F(t, 0)) \leq \ell(t) \quad \text{a.e.}.$$

Then
1) for any $\alpha > 1$, every solution $x(t)$ of M) satisfies $\int_0^\infty e^{-\alpha L(t)} \|x'(t)\| \, dt < \infty$, $L(t) \equiv \int_0^t \ell(s) \, ds$;
2) for every $\alpha > 1$, the map $x_0 \mapsto S(x_0)$ is Lipschitz continuous from X to 2^X. The values of this solution map are nonempty closed subsets of the Banach space

$$X_\alpha = \left\{ x(\cdot) \mid x \colon [0, \infty) \to X,\ \|x\| = x(0) + \int_0^\infty e^{-\alpha L(t)} \|x'(t)\| \, dt \text{ exists} \right\}.$$

This result is proved by applying Lim's theorem to the mapping T defined on X_α by:

$$T \colon u(\cdot) \longmapsto \left\{ \varphi(\cdot) \mid \varphi(t) \in F\left(t, x_0 + \int_0^t u(s) \, ds\right),\ \int_0^\infty e^{-\alpha L(t)} \|\varphi(t)\| \, dt < \infty \right\}.$$

A function $x(t)$ will solve M) if and only if $x'(t)$ is a fixed point of T. By creating a map on $x'(t)$ ($= u(t)$), Kánnai and Tallos can use integral operators without introducing mild solutions. They also give an elegant proof of a multifunction version of Gronwall's inequality:

Theorem 5.12. Let X and $F(t, x)$ be as in Theorem 5.11. Suppose $y(t)$ is an absolutely continuous function from $[0, \infty)$ to X satisfying

$$\tilde{h}(y'(t), F(t, y(t))) \leq p(t) \quad \text{for a.e. } t \in [0, \infty),$$

with p locally integrable. Then for any $\alpha > 1$ there exists a solution $x_\alpha(t)$ of M) such that for all $t \in (0, \infty)$

$$\|x_\alpha(t) - y(t)\| \leq \frac{2\alpha + 1}{\alpha - 1} e^{\alpha L(t)} \left(\|x_0 - y(0)\| + \int_0^\infty e^{-\alpha L(s)} p(s) \, ds \right).$$

Couchouran and Kamenskii [30] consider a problem motivated by parabolic differential inclusions:

$$\left(\frac{dx}{dt} - Ax \right) \in f(t, x, \theta),\ t \in [0, T],\ \theta \in [0, 1],\ x(0) = x_0, \qquad M^*_\theta)$$

with $x(t) \in X$, X a separable Banach space, f taking closed compact nonempty values, A linear with dense domain $D(A)$. They assume that A is accretive in a generalized sense, and that f is condensing in a generalized sense. Under some additional technical, but relatively mild, asuumptions they prove that if

$$\theta_n \to \theta, \quad x_0^{(n)} \to x_0 \quad \text{as} \quad n \to \infty,$$

then
1) $M^*_{\theta_n}$ has at least one mild solution $x_n(t)$, $x_n(0) = x_0^{(n)}$;
2) if $\{x_n(\cdot)\}$ is any sequence of such solutions, this sequence is relatively compact in $C([0,T], X)$;
3) any cluster point of such a sequence is a solution of M^*_θ).

Our final theorem describes a recent result for non-convex differential inclusions.

Theorem 5.13. (DeBlasi, Pianigiani and Staicu [39]) *Consider the problem*

$$\dot{x}(t) = Ax(t) + F(t, x(t), \eta), \quad x(0) = \xi \tag{D}$$

for $t \in [0, a]$, $x(t) \in E$ a separable Banach space, $\eta \in H$ a separable metric space. Assume that A is the infinitesimal generator of a C^0 semigroup of bounded operators $T(t)$, $t \geq 0$, on E. Assume that F is a measurable, Hausdorff continuous in η, closed-valued multifunction which satisfies

$$h\big(F(t, x, \eta), F(t, y, \eta)\big) \leq \ell(t)|x - y|$$
$$h\big(0, F(t, x, y)\big) \leq k(t)$$

with ℓ and k in $L^1[0, a]$.

Then the set $S(\xi, \eta)$ of all mild solutions of (D) is a retract of a convex subset of a Banach space. This retraction on S can be taken to be continuous in (ξ, η). Thus S is contractible in itself and also $(\xi, \eta) \multimap S(\xi, \eta)$ admits a continuous selection. Any two such selections can be joined by a homotopy in $S(\xi, \eta)$.

6. Boundary Value Problems

Our present understanding of the structure of the solution set for boundary value problems involving differential equations in abstract spaces is very incomplete. Almost all of the few known existence results date from the 1980's, and little has been proved about the topological structure of the solution set.

Nevertheless, in the finite dimensional case this subject has been studied by various mathematicians, who have among other things proved Aronszajn type theorems. We mention Anichini and G. Conti [4], Nieto (see [97], [98] and [99]) and Kannan *et al.* [73] for the Dirichelet problem, Nieto and Sanchez [100] for the Duffing equation, Nieto [101] for a Sturm-Liouville problem, Bielawski and Pruszko [12] for the Floquet and Nicoletti problems, Vidossich [153] for the terminal value problem, Gupta, Nieto and Sanchez [61] and Bebernes and Martelli [9] for other boundary value problems. Peixoto and Thom [112] established results on dimension and smoothness of certain manifolds related to solution sets of two point boundary value problems for second order nonlinear equations.

Let X be a Banach space, $J \equiv [a,b] \subset \mathbb{R}$ and $f(t,x,y)$ a continuous function on $J \times X^2$. Consider the following Sturm-Liouville problem:

$$\begin{cases} \ddot{x}(t) = f(t, x(t), \dot{x}(t)) \\ c_1 x(a) - d_1 \dot{x}(a) = x_1 \\ c_2 x(b) + d_2 \dot{x}(b) = x_2 \end{cases} \quad \text{BVP)}$$

where $c_i, d_i \in \mathbb{R}$ and $x_i \in X$ for $i = 1, 2$.

We are interested only in the so-called incompatible case, that is we shall suppose the existence of a Green's function for BVP). In fact a counterexample in ℓ_2 constructed by Deimling [40] shows that in the compatible case existence can fail even if the function f is compact and globally defined and globally continuous.

Recall that a Green's function for the problem BVP) is a continuous scalar function $G(t,s)$ such that, for any fixed function $h \in C(J, X)$, the unique solution of the problem:

$$\begin{cases} \ddot{x}(t) = h(t) \\ c_1 x(a) - d_1 \dot{x}(a) = 0 \\ c_2 x(b) + d_2 \dot{x}(b) = 0 \end{cases}$$

can be expressed as $x(t) = \int_a^b G(t,s)h(s)ds + g(t)$ where $g \in C^2(J, X)$ is the unique solution of

$$\begin{cases} \ddot{x}(t) = 0 \\ c_1 x(a) - d_1 \dot{x}(a) = x_1 \\ c_2 x(b) + d_2 \dot{x}(b) = x_2. \end{cases}$$

It is known (see Bernfeld and Lakshmikantham [11]) that when a Green's function exists, the problem BVP) is equivalent to the integral equation:

$$x(t) = \int_a^b G(t,s) f\big(s, x(s), \dot{x}(s)\big) ds + g(t).$$

From now on, V will denote the class of functions $\omega : J \times \mathbb{R}^2 \to \mathbb{R}$, linear on $\mathbb{R}^2$, such that $\omega(t, \varphi(t), \psi(t)) \geq 0$ when both φ and ψ are nonnegative on all of J, and the linear operator H from $C(J, \mathbb{R}) \times C(J, \mathbb{R})$ into itself, defined $\forall \varphi, \psi \in C(J, \mathbb{R})$ as:

$$H(\varphi, \psi) = \left(\int_a^b |G(t,s)| \omega(s, \varphi(s), \psi(s)) ds, \int_a^b |G_t(t,s)| \omega(s, \varphi(s), \psi(s)) ds \right)$$

has spectral radius $r(H) < 1$.

The set V has the following property:

Theorem 6.1. *Let $\omega \in V$ and $\varphi, \psi \in C(J, \mathbb{R})$ such that:*

$$\varphi(t) \leq \int_a^b |G(t,s)| \omega\big(s, \varphi(s), \psi(s)\big) ds \quad \text{and}$$

$$\psi(t) \leq \int_a^b |G_t(t,s)| \omega\big(s, \varphi(s), \psi(s)\big) ds.$$

Then we have $\varphi, \psi \leq 0$ on J.

Proof. By hypothesis we have $(\varphi,\psi) \le H(\varphi,\psi)$ on J, that is $(Id-H)(\varphi,\psi) \le (0,0)$. As the spectral radius of H is less than 1, the operator $(Id-H)^{-1}$ exists, is linear and continuous, and has the following representation (see [76]):

$$(Id - H)^{-1} = \sum_{n=1}^{\infty} H^n.$$

From the properties of the integral and since ω is linear and positive, it follows immediately that the inequality $(f,g) \le (0,0)$ implies $H(f,g) \le (0,0)$ $\forall\, f,\ g \in C(J,\mathbb{R})$ on J. By induction it is easy to see that this relation is satisfied by H^n $\forall\, n \in N$ and so by $(Id-H)^{-1}$. Consequently, we obtain:

$$(\varphi,\psi) \le (Id-H)^{-1}(Id-H)(\varphi,\psi) \le (0,0) \quad \text{on} \quad J.$$

q.e.d

We will need the following theorem on interchanging integration and the measure of noncompactness ([1]).

Theorem 6.2. *Let X be a wcg space and $A = \{x_n\}_{n \in N} \subset C(J,X)$ a bounded sequence. Then $\varphi(t) = \chi(A(t))$ is measurable $\forall\, t \in J$ and we have:*

$$\chi\left(\left\{\int_a^b x_n(s)ds\right\}_{n \in N}\right) \le \int_a^b \varphi(s)ds.$$

In order to show the existence of solutions for the problem BVP), we shall use a particular version of a fixed point theorem due to Sadovskii ([123], [1]):

Theorem 6.3. *Let X be a Banach space, $K \subset X$ a closed and convex set and $F : K \to K$ a continuous operator. Suppose that for some $x \in K$, F satisfies the following condition:*

if $D \subset X$ is countable and $\overline{D} = \operatorname{co}(\{x\} \cup F(D))$ then D is precompact. a)

Then F has at least one fixed point in K.

The next theorem, in which the existence part is due to Monch [91], is one of the earliest results concerning the topological structure of the solution set of a boundary value problem in abstract spaces.

Theorem 6.4. Let X be a wcg space, $J \equiv [a,b]$ and for $i = 1,2$ let $d_i \in \{0,1\}$, $c_i \in \mathbb{R}$ such that if $d_i = 0$ then $c_i = 1$. Assume that $f : J \times X^2 \to X$ is a continuous function satisfying the following conditions:

b$_1$) $\|f(t,x,y)\| \leq M \quad \forall\, x, y \in X$
b$_2$) $\chi(f(t, A_1, A_2)) \leq \omega(t, \chi(A_1), \chi(A_2)) \quad \forall t \in J \;\; \forall A_1, A_2$ bounded sets in X and $\omega \in V$.

Then the solution set S of BVP) is nonempty and compact.

Proof. Define the operator $T : C^1(J, X) \to C^1(J, X)$ as:

$$T(x)(t) = \int_a^b G(t,s) f(s, x(s), \dot{x}(s)) \, ds + g(t)$$

where $g \in C^2(J, X)$ is the solution of $\ddot{x}(t) = 0$ described above the statement of Theorem 6.1, and $G(t, s)$ is the Green's function for BVP). Obviously the set S and the set of fixed points of T coincide on $C^1(J, X)$. Let m_1, m_2 and m_3 be three positive constants such that

$$\int_a^b |G(t,s)| ds \leq m_1 \qquad \int_a^b |G_t(t,s)| ds \leq m_2 \qquad \|g_t(t)\| \leq m_3.$$

For the exact expressions of m_1, m_2, m_3 as functions of a, b, c_i, d_i, x_i see Chandra, Lakshmikantham and Mitchell [25] or Bernfeld and Lakshmikantham [11].

Given $\varepsilon > 0$, by the continuity of f there exists δ_ε such that $\forall\, x_1, z_1, x_2, z_2 \in X$ with $\|x_1 - x_2\| + \|z_1 - z_2\| < \delta_\varepsilon$ we have $\|f(t, x_1, z_1) - f(t, x_2, z_2)\| < \varepsilon$. Now let $x, y \in C^1(J, X)$ with $\|x - y\|_1 = \|x - y\| + \|\dot{x} - \dot{y}\| < \delta_\varepsilon$. Then:

$$\left\| \int_a^b G(t,s) \big(f(s, x(s), \dot{x}(s)) - f(s, y(s), \dot{y}(s)) \big) ds \right\|$$

$$+ \left\| \int_a^b G_t(t,s) \big(f(s, x(s), \dot{x}(s)) - f(s, y(s), \dot{y}(s)) \big) ds \right\|$$

$$\leq \int_a^b |G(t,s)| \cdot \| f(s, x(s), \dot{x}(s)) - f(s, y(s), \dot{y}(s)) \| ds$$

$$+ \int_a^b |G_t(t,s)| \cdot \| f(s, x(s), \dot{x}(s)) - f(s, y(s), \dot{y}(s)) \| ds$$

$$\leq (m_1 + m_2)\varepsilon,$$

$$\|Tx - Ty\|_1 = \sup_{t \in [a,b]} \|(Tx - Ty)(t)\| + \sup_{t \in [a,b]} \left\|\frac{d}{dt}(Tx - Ty)(t)\right\| \leq (m_1 + m_2)\varepsilon.$$

Consequently the operator T is continuous on the space $C^1(J,X)$ which is closed and convex with the norm $\|\cdot\|_1$. In order to apply Theorem 6.3, we need to verify property a). To this end consider a sequence $D = \{x_n\}_{n\in N} \subset C^1(J,X)$ with $\overline{D} = \overline{co}\,(\{x\} \cup T(D))$ for some fixed x. By the properties of measure of noncompactness we obtain:

$$\chi(D) = \chi(\overline{D}) = \chi(\overline{co}\,(\{x\} \cup T(D))) = \chi(\{x\} \cup T(D)) = \chi(T(D)) \qquad *)$$

that is, and scale D is precompact if and only if $T(D)$ is precompact.

Given $z \in T(D)$, hypothesis b_1) implies that there exists $y \in X$ such that $\|\ddot{z}(t)\| = \|f(t,y(t),\dot{y}(t))\| \leq M$ and so

$$\|\dot{z}(t)\| = \left\|\int_a^b G_t(t,s)(f(s,y(s),\dot{y}(s))\,ds\right\| + \|g_t(t)\| \leq m_2 M + m_3.$$

Applying the mean-value theorem, $\forall\, t_1, t_2 \in J$ we obtain:

$$\|z(t_1) - z(t_2)\| \leq (m_2 M + m_3)|t_1 - t_2| \quad \text{and}$$

$$\|\dot{z}(t_1) - \dot{z}(t_2)\| \leq M|t_1 - t_2|, \quad \forall\, z \in T(D).$$

It follows that the sets $T(D)$ and $\dot{T}(D)$ are equicontinuous.

Now define two functions $\varphi, \psi : \mathbb{R} \to \mathbb{R}^+$ as:

$$\varphi(t) = \chi(T(D)(t)) \quad \text{and} \quad \psi(t) = \chi(\dot{T}(D)(t)).$$

From $*)$ we have $\varphi(t) = \chi(D(t))$ and $\psi(t) = \chi(\dot{D}(t))$. The first is obvious, the second follows from the fact that if U and V are sets in $C^1(J,X)$ with $\overline{U} = \overline{V}$, then $\overline{\dot{U}} = \overline{\dot{V}}$ in $C(J,X)$. Theorem 6.2 and the condition b_2) imply the following relations:

$$\varphi(t) = \chi\left(\left\{\int_a^b G(t,s)f(s,x_n(s),\dot{x}_n(s))\,ds + g(t)\right\}_{n\in N}\right)$$

$$= \chi\left(\left\{\int_a^b G(t,s)f(s,x_n(s),\dot{x}_n(s))\,ds\right\}_{n\in N}\right)$$

$$\leq \int_a^b \chi(\{G(t,s)f(s,x_n(s),\dot{x}_n(s))\}_{n\in N})ds$$

$$= \int_a^b |G(t,s)|\chi(\{f(s,x_n(s),\dot{x}_n(s))\}_{n\in N})ds$$

$$\leq \int_a^b |G(t,s)|\omega(s,\chi(D(s)),\chi(\dot{D}(s)))ds$$

$$= \int_a^b |G(t,s)|\omega(s,\varphi(s),\psi(s))ds \quad \forall t \in J.$$

Similarly we obtain $\psi(t) \leq \int_a^b |G_t(t,s)|\omega(s,\varphi(s),\psi(s))ds$. As the functions φ and ψ are nonnegative, Theorem 6.1 implies $\varphi \equiv \psi \equiv 0$ on J. From the Ascoli-Arzelà theorem on $C^1(J,X)$ it follows that $T(D)$ is precompact, and so the problem BVP) admits at least one solution.

Now we prove that S is closed in $C^1(J,X)$. Consider the sequence $\{x_n\}_{n\in N} \subset S$ with $\lim_{n\to\infty} x_n = x_0$ in $C^1(J,X)$.

Clearly $x_0 \in C^1(J,X)$, moreover the continuity of f implies $\lim_{n\to\infty} f(t,x_n(t),\dot{x}_n(t)) = f(t,x_0(t),\dot{x}_0(t))$. Hence:

$$x_0(t) = \lim_{n\to\infty} x_n(t) = \lim_{n\to\infty} \int_a^b G(t,s)f(t,x_n(s),\dot{x}_n(s))ds + g(t)$$

$$= \int_a^b G(t,s)f(t,x_0(s),\dot{x}_0(s))ds + g(t)$$

that is, $x_0 \in S$.

Finally we show that S is precompact in $C^1(J,X)$. Given a sequence $A = \{y_n\}_{n\in N} \subset S$, we introduce two functions $\phi(t) = \chi(A(t))$ and $\sigma(t) = \chi(\dot{A}(t))$. Noting that $\chi(T(A)) = \chi(A)$ we have $\phi(t) = \chi(T(A)(t))$ and $\sigma(t) = \chi(\dot{T}(A)(t))$. Therefore, with identical arguments to those used for the sequence D, we see that A and $\dot{A}$ are equicontinuous and $A(t)$ is precompact $\forall t \in J$. Once more, the Ascoli-Arzelà theorem in $C^1(J,X)$ implies the compactness of S.

q.e.d.

We note that the existence results obtained by Chandra, Lakshmikantham and Mitchell [25] and Deimling [44] are particular cases of Theorem 6.4. In their papers the condition b_2) is replaced respectively by the stronger conditions:

$$\alpha(f(J, A, B)) \le k \cdot \max\{\alpha(A), \alpha(B)\} \qquad \text{d)}$$

$$\alpha(f(t, A, B)) \le k(\alpha(A) + \alpha(B)) \qquad \text{c)}$$

where k is a constant sufficiently small to make automatically true the hypotheses of Theorem 6.1. The results in [25] and [44] in turn contain an existence theorem due to Schmitt and Volkmann [124] in which the compactness of f is assumed, that is $\alpha(f(t, A, B)) = 0$.

We now consider the multivalued Sturm-Liouville problem, which has been recently studied by Deimling [46]. With the same notation as was used for the single-valued problem, we consider the following:

$$\begin{cases} \ddot{x}(t) \in F(t, x(t), \dot{x}(t)) \\ c_1 x(a) - d_1 \dot{x}(a) = x_1 \\ c_2 x(b) + d_2 \dot{x}(b) = x_2 \end{cases} \qquad \text{BVM)}$$

where $F : J \times X^2 \to \mathcal{P}(X)$ and $\mathcal{P}(X)$ is the power set of X without the empty set.

This time the problem BVM) is equivalent to the integral inclusion $x \in Tx$ where the multivalued map $T : C^1(J, X) \to \mathcal{P}(C^1(J, X))$ is defined as:

$$Tx = g + \left\{ y \in C^1(J, X) : \right.$$

$$\left. y(t) = \int_a^b G(t, s) v(s) ds \text{ on } J \text{ and } v(\cdot) \in F(\cdot, x(\cdot), \dot{x}(\cdot)), \ v \text{ measurable} \right\}.$$

In order to prove the existence of solutions of the problem BVM), we shall use a fixed point theorem for multifunctions due to F. Browder (for a full discussion, see Deimling [46], §11).

Theorem 6.5. *Let X be a Banach space, $D \subset X$ a compact and convex set and $T : D \multimap X$ an u.s.c. multifunction with closed and convex values. Suppose that $Tx \cap \bar{I}_D \ne \emptyset \quad \forall\, x \in D$ where $\bar{I}_{D(x)} = \{(1 - \lambda)x + \lambda y : \lambda \ge 0, \text{ and } y \in D\}$. Then T admits a fixed point.*

Concerning selections of a multivalued map, we have the following:

Theorem 6.6. *Let X be a Banach space, $J \equiv [a,b]$, $v \in C(J,X)$ and $F: J \times X \multimap X$ with compact values. Assume that for almost all $t \in J$, $F(t,\cdot)$ is u.s.c., and $\forall\, x\ F(\cdot\,,x)$ has a strongly measurable selection. Then there exists a measurable $w(\cdot) \in F(\cdot\,,v(\cdot))$.*

Using this result, we can prove the following result due to Deimling:

Theorem 6.7. *Let X be a separable Banach space, $J \equiv [a,b] \subset \mathbb{R}$ and $F: J \times X^2 \multimap X$ a multifunction with closed and convex values satisfying the following conditions:*

i) $\forall\, x, y\ F(\cdot, x, y)$ has a strongly measurable selection and $\forall\, t \in J\ \ F(t, \cdot, \cdot)$ is u.s.c.,

ii) There exist three scalar functions $k, \ell, m \in C^1(J, R)$ such that
$\|F(t,x,y)\| \leq k(t)\|x\| + \ell(t)\|y\| + m(t)$ on $J \times X^2$,

iii) $\chi(F(t, A, D)) \leq k(t)\chi(A) + \ell(t)\chi(D)$ on $J\quad \forall\, A, D$ bounded sets of X.

iv) the linear operator H defined as:

$$H(\varphi, \psi) = \left(\int_a^b |G(t,s)|(k(s)\varphi(s) + \ell(s)\psi(s))\,ds, \right.$$

$$\left. \int_a^b |G_t(t,s)|(k(s)\varphi(s) + \ell(s)\psi(s))\,ds \right)$$

has spectral radius $r(H) < 1$.

Then the problem BVM) admits a solution.

Proof. At first we prove the boundedness in $C^1(J, X)$ of the (possibly empty) set S of fixed points of T. Let $x \in S$, that is $x \in Tx$. Then we have $x \in g + \int_a^b G(\cdot.s)F(s, x(s), \dot{x}(s))\,ds$ and:

$$\|x(t)\| \leq \|g(t)\| + \int_a^b |G(t,s)|\,\|k(s)x(s) + \ell(s)\dot{x}(s) + m(s)\|ds$$

$$= \varphi_0(t) + \int_a^b |G(t,s)|(k(s)\|x(s)\| + \ell(s)\|\dot{x}(s)\|)\,ds \quad \text{where}$$

$$\varphi_0(t) = \|g(t)\| + \int_a^b |G(t,s)|m(s)ds.$$

Similarly, setting $\psi_0(t) = \|g(t)\| + \int_a^b |G_t(t,s)| m(s) ds$, we obtain:

$$\|\dot{x}(t)\| \leq \psi_0(t) + \int_a^b |G_t(t,s)|(k(s)\|x(s)\| + \ell(s)\|\dot{x}(s)\|) ds.$$

This implies $(Id - H)\left(\|x(\cdot)\|, \|\dot{x}(\cdot)\|\right) \leq (\varphi_0, \psi_0)$ on J.

From the hypothesis $r(H) < 1$ we have $(Id - H)^{-1} = \sum_{n=1}^{\infty} H^n$ and, since H^n is monotone with respect to the relation $\leq$, the linear continuous operator $(Id - H)^{-1}$ also has this property. Consequently, applying $(Id - H)^{-1}$ to both the sides of the preceding inequality, we see that:

$$(\|x(\cdot)\|, \|\dot{x}(\cdot)\|) \leq (Id - H)^{-1}(\varphi_0, \psi_0) \quad \text{on} \quad J.$$

From the boundedness of $(Id - H)^{-1}$ the boundedness of S in $C^1(J, X)$ follows immediately.

Now consider the set $D = \{x \in C^1(J, X) : (\|x(\cdot)\|, \|\dot{x}(\cdot)\|) \leq (Id - H)^{-1}(\varphi_0, \psi_0)\}$. D is bounded, closed because of the sign $\leq$, and contains the fixed points of T. Moreover, given x and $y \in D$, $\forall \lambda \in [0, 1]$ we have:

$$(\|\lambda x + (1-\lambda)y\|, \|\lambda \dot{x} + (1-\lambda)\dot{y}\|)$$
$$\leq (\lambda\|x\| + (1-\lambda)\|y\|, \lambda\|\dot{x}\| + (1-\lambda)\|\dot{y}\|$$
$$= \lambda(\|x\|, \|\dot{x}\|) + (1-\lambda)(\|y\|, \|\dot{y}\|)$$
$$\leq \lambda(Id - H)^{-1}(\varphi_0, \psi_0) + (1-\lambda)(Id - H)^{-1}(\varphi_0, \psi_0)$$
$$= (Id - H)^{-1}(\varphi_0, \psi_0).$$

Therefore D is convex. Since $(\varphi_0, \psi_0) \geq (0, 0)$ on J, we have:

$$(Id - H)^{-1}(\varphi_0, \psi_0) \geq (Id - H)^{-1}(0, 0) = (0, 0) \quad \text{that is} \quad 0 \in D.$$

Let $\bar{x} \in T(D)$. Then $\bar{x} \in Tx$ with $x \in D$ and :

$$\|\bar{x}(t)\| \leq \varphi_0(t) + \int_a^b |G(t,s)|(k(s)\|x(s)\| + \ell(s)\|\dot{x}(s)\|) ds$$

$$\|\dot{\bar{x}}(t)\| \leq \psi_0(t) + \int_a^b |G_t(t,s)|(k(s)\|x(s)\| + \ell(s)\|\dot{x}(x)\|) ds$$

that is

$$(\|\bar{x}(\cdot)\|, \|\dot{\bar{x}}\|) \leq H(\|x(\cdot)\|, \|\dot{x}(\cdot)\|) + (\varphi_0, \psi_0)$$
$$\leq H(Id - H)^{-1}(\varphi_0, \psi_0) + (\varphi_0, \psi_0)$$
$$= (H(Id - H)^{-1} + Id)(\varphi_0, \psi_0) = (Id - H)^{-1}(\varphi_0, \psi_0).$$

So we have $\bar{x} \in D$ and $T(D) \subset D$. Therefore $T(D)$ is bounded and $\forall x \in T(D)$ the functions x, $\dot{x}$ are bounded in norm on J. Now the mean-value theorem immediately implies the equicontinuity of $T(D)$ in $C^1(J, X)$, which we can express as:

$$\|x(t) - x(s)\| + \|\dot{x}(t) - \dot{x}(s)\| \leq d(|t-s|) \quad \text{with} \quad (t,s) \in J \times J$$

where $d : \mathbb{R}^+ \to \mathbb{R}^+$ is a continuous function such that $\lim_{t \to 0^+} d(t) = 0$.

Now observe that the values of F are necessarily compact. In fact from condition iii) and from the properties of χ we have

$$\chi(F(t, x, y)) \leq k(t)\chi(\{x\}) + \ell(t)\chi(\{y\}) = 0 \quad \forall x, y \in X.$$

Therefore F satisfies the hypotheses of Theorem 6.6, which implies $Tx \neq \emptyset \quad \forall x \in D$. Let $x_1, x_2 \in Tx$ that is $x_1 = g + \int_a^b G(\cdot, s)v_1(s)ds$ and $x_2 = g + \int_a^b G(\cdot, s)v_2(s)ds$ with $v_1, v_2 \in F(s, x, \dot{x})$. Recalling that the values of F are convex, we obtain:

$$\alpha x_1 + (1-\alpha)x_2 = g + \int_a^b G(\cdot, s)(\alpha v_1 + (1-\alpha)v_2)(s)\, ds$$

$$= g + \int_a^b G(\cdot, s)w_3(s)\, ds$$

where $w_3 \in F(s, x, \dot{x})$. Consequently the mapping T has convex values.

Now we introduce the sequence of sets $\{D_n\}_{n \in N}$ defined as:

$$D_0 = \{0\} \qquad D_{n+1} = \operatorname{co}\left(\{0\} \cup T(D_n)\right).$$

Let $V = \bigcup_{n=1}^{\infty} D_n$. As $0 \in D$ and $T(D) \subset D$, $\forall n \in N$ we have $D_n \subset D$. Since the sequence $\{D_n\}_{n \in N}$ is increasing, the following equality holds:

$$V = \operatorname{co}\left(\{0\} \cup T(V)\right). \qquad *)$$

So $\overline{V} \in D$ is closed, convex and bounded. As $T(V) \subseteq T(D)$, $T(V)$ is also equicontinuous in $C^1(J, X)$. Thus, by the Ascoli-Arzelà theorem, the set V will be precompact if we succeed in proving the $V(t) = \{x(t) : x \in V\}$ is precompact in X $\forall t \in J$. Since $\chi(V(t)) = \chi(T(V)(t))$, we need only show the precompactness of $T(V)(t)$ $\forall t \in J$. Let $\{x_n\}_{n \in N} \subset T(V)$ and $\varphi : J \to \mathbb{R}^+$ such that:

$$\varphi(t) = \chi(\{x_n\}_{n \in N}) \quad \text{and}$$

$$x_n = g + \int_a^b G(\cdot, s) w_n(s) ds \quad \text{with} \quad w_n(\cdot) \in F(\cdot, v_n(\cdot), \dot{v}(\cdot))$$

for some bounded sequence $\{v_n\}_{n \in N} \subset V$. Theorem 6.2 together with the condition iii) imply:

$$\varphi(t) \leq \chi\left(\left\{\int_a^b G(t,s) F(s, v_n(s), \dot{v}_n(s)) ds\right\}_{n \in N}\right)$$

$$\leq \int_a^b |G(t,s)| \big(k(s)\chi(\{v_n(s)\}_{n \in N}) + \ell(s)\chi(\{\dot{v}_n(s)\}_{n \in N})\big) ds.$$

Now consider the following function $\omega : \mathcal{P}(X) \to \mathbb{R}^+$:

$$\omega(B) = \sup \{\chi(\{x_n\}_{n \in N}) : \{x_n\}_{n \in N} \subset B\} \quad \text{with} \quad B \subset X \quad \text{a bounded set.}$$

As V is equicontinuous, the functions $\omega(V(\cdot))$ and $\omega(\dot{V}(\cdot))$ are continuous. Hence:

$$\omega(V(t)) \leq \int_a^b |G(t,s)| \big(k(s)\omega(V(s)) + \ell(s)\omega(\dot{V}(s))\big) ds$$

and analogously

$$\omega(\dot{V}(t)) \leq \int_a^b |G_t(t,s)| \big(k(s)\omega(V(s)) + \ell(s)\omega(\dot{V}(s))\big) ds$$

so that $(\omega(V(\cdot)), \omega(\dot{V}(\cdot))) \leq (I - H)^{-1}(0, 0) = (0, 0)$ and $\chi(V(\cdot)) = \chi(\dot{V}(\cdot)) = 0$ on J.

So far we know that the values of T on D are nonempty, convex and bounded; let us show that they are compact. Let $x \in D$ and $\{x_n\}_{n \in N} \subset Tx$. Then:

$$x_n = g + \int_a^b G(\cdot, s) w_n(s) ds \quad \text{with} \quad w_n(s) \in F(s, x(s), \dot{x}(s)).$$

But $F(s, x(s), \dot{x}(s))$ is compact, so that there exists a subsequence $w_{n_k}(s)$ converging to $w_0(s) \in F(s, x(s), \dot{x}(s))$. Since $\lim_{k \to \infty} w_{n_k}(s) = w_0(s)$ pointwise on $[a, b]$, and $\|w_{n_k}(s)\| \leq \|F(s, x(s), \dot{x}(s))\| \leq k(s)\|x\| + \ell(t)\|\dot{x}\| \in L^1(J, \mathbb{R}^+)$, Lebesgue's dominated convergence theorem implies:

$$\lim_{k \to \infty} x_{n_k} = \lim_{k \to \infty} g + \int_a^b G(\cdot, s) w_{n_k}(s) ds = g + \int_a^b G(\cdot, s) w_0(s) ds \in Tx$$

so Tx is compact.

Therefore $T : \overline{V} \to \mathcal{P}(D)$ is an u.s.c. multifunction with compact and convex values on the compact set $\overline{V}$. Let $x_0 \in \overline{V}$ and $\{x_n\}_{n \in N} \subset V$ be such that $x_n \to x_0$. Then from relation $*$) it follows that:

$$Tx_n \subset T(V) \subset V \subset \overline{V} \quad \forall \, n \in N.$$

From the properties of u.s.c. multifunctions we obtain:

$$\sup_{x \in \overline{V}} \tilde{h}(x, Tx_0) \leq \sup_{x \in Tx_n} \tilde{h}(x, Tx_0) \to 0 \quad \text{as} \quad n \to \infty$$

so $\sup_{x \in \overline{V}} \tilde{h}(x, Tx_0) = 0$. But $\overline{V}$ and Tx_0 are compact, so this last equality implies $Tx_0 \cap \overline{V} \neq \emptyset$. Observe that $\overline{V}$ is also convex, which implies $\overline{V} \subset \overline{I}_D(x_0)$ and $Tx_0 \cap \overline{I}_D(x_0) \neq \emptyset \quad \forall \, x_0 \in \overline{V}$. Now Theorem 6.5 assures the existence of fixed points of T.

q.e.d.

Notice that in the presence of single-valued mappings this theorem becomes a particular case of Theorem 6.4. In fact a more general estimate for f is used there, involving a function $\omega(t, \chi(A), \chi(B))$ such that $\omega(\cdot, x, y)$ may not be integrable, contrary to Hypothesis iii) of Theorem 6.7.

We conclude this chapter by discussing some important results concerning the structure of the solution set of other kinds of equations.

For functional differential equations we mention the Kneser type theorems obtained by Bulgakov and Lyapin [20], Sosulski [128], Haddad [62], Krbec and Kurzweil [81], Lasry and Robert ([87] and [88]) in $\mathbb{R}^n$ and Shin [126] in a Banach space. This last deals with the following problem:

$$\begin{cases} \dot{x}(t) = f(t, x_t) \\ x_\sigma = \varphi. \end{cases}$$

The method of proof is similar to that of Theorem 3.8.

The main results for integral equations are the Kneser type theorems due to Kelley [76] for Volterra equations, Bulgakov and Lyapin ([18] and [19]) and Szufla [141] for Hammerstein equations and the Aronszajn theorem obtained by Szufla [136] for Volterra equations.

Concerning parabolic partial differential equations we mention the articles of Talaga ([145] and [146]) and Bebernes and Schmitt [10] (Kneser type theorems) and Nieto ([94] and [102]) and Ballotti [8] (Aronszajn type theorems).

The elliptic case was studied by Nieto ([95] and [96]) who proved that the solution set is acyclic and compact.

Finally, Aronszajn type theorems about the following Darboux problem for hyperbolic equations:

$$\begin{cases} \dfrac{\partial^2 z}{\partial x \partial y} = f(x,y,z), & (x,y) \in K, \\ z(x,0) = 0, & 0 \le x \le d_1, \\ z(0,y) = 0, & 0 \le y \le d_2, \end{cases}$$

are obtained in $\mathbb{R}^n$ by De Blasi and Myjak [36], Gorniewicz and Pruszko [58], Gorniewicz, Bryszewski and Pruszko [59]. Very recently Szufla and Bugajewski [144] have proved that in a Banach space context the set of weak solutions is a continuum. Margheri and Zecca [90] have extended results proved by De Blasi and Pianigiani [38] for finite dimensional spaces. They give conditions under which the solution set BVM) is a compact retract of either $W^{2,1}(J,X)$ or $C^1(J,X)$. They define a function $F: J \times X^2 \to \mathcal{P}(X) \setminus \phi$ to be *almost lower semicontinuous* if

$$\forall \varepsilon > 0, \quad \forall K \subset X^2 \text{ nonempty compact}, \quad \exists \text{ closed } J_\varepsilon \subset J$$

with $\mu(J \setminus J_\varepsilon) < \varepsilon$ such that $F|_{J_\varepsilon \times K}$ is $\ell.s.c.$ and

$$\text{span } F(J_\varepsilon \times K) \text{ is separable.}$$

Theorem 6.8. *Let* $F: J \times X^2 \to \mathcal{P}(X) \setminus \phi$ *be almost* $\ell.s.c.$ *with compact values. Assume that* $L^1(J,X)$ *is separable and that there exist* $m \in L^1(J,X)$, *and constants* $\mu, \lambda > 0$ *such that*
1) $F(\cdot, x, y)$ *is measurable* $\forall (x,y) \in X^2$,
2) $h(F(t,0,0), \{0\}) \le m(t)$ *a.e. on* J;

3) $h(F(t,x_1,y_1), F(t,x_2,y_2)) \leq \mu\|x_1 - x_2\| + \lambda\|y_1 - y_2\|$
$\forall\, (t, x_i, y_i) \in J \times X \times X$, with

$$\mu\|G(t,s)\|_\infty + \lambda\|G_t(t,s)\|_\infty < 1.$$

Then the set of solutions of BVM) is a retract of $W^{2,1}(J,X)$.

If in addition we assume that F has convex values, that X is separable and reflexive, and that

$$m \in L^2(J,X), \quad (b-a)(\mu + \lambda)\, \max\{\|G(t,s)\|_\infty, \|G_t(t,s)\|_\infty\} < 1,$$

then the solution set is a retract of $C^1(J,X)$.

7. Bibliography

[1] R. R. Akhmerov, M. I. Kamenskii, A. S. Potapov, A. E. Rodkina and B. N. Sadovskii, *Measures of Noncompactness and Condensing Operators* (translated from Russian), Birkhauser, Berlin, 1992.

[2] A. Ambrosetti, *Un teorema di esistenza per le equazioni differenziali negli spazi di Banach*, Rend. Sem. Mat. Univ. Padova, **39** (1967), 349–360.

[3] G. Anichini and P. Zecca, *Multivalued differential equations in Banach space: an application to control theory*, J. Optim. Th. Appl. **21** (1977), 477–486.

[4] G. Anichini and G. Conti, *Existence of solutions of a boundary problem through the solution map of linearized type problem*, Sem. Mat. Univ. Torino.

[5] M. Aronszajn, *Le correspondent topologique de l'unicité dans le théorie des équations différentielles*, Ann. Math. **43** (1942), 730–738.

[6] Z. Artstein, *Continuous dependence on parameters of solutions of operator equations*, Trans. Amer. Math. Soc. **231** (1977), 143-166.

[7] J. P. Aubin and A. Cellina, *Differential Inclusions*, Springer-Verlag, Berlin, 1984.

[8] M E. Ballotti, *Aronszajn's theorem for a parabolic partial differential equation*, Nonl. Anal. T.M.A. **9**, No. 11 (1985), 1183–1187.

[9] J. Bebernes and M. Martelli, *On the structure of the solution set for periodic boundary value problems*, Nonl. Anal. T.M.A. **4**, No. 4 (1980), 821–830.

[10] J. Bebernes and K. Schmitt, *Invariant sets and the Hukuhara-Kneser property for systems of parabolic partial differential equations*, Rocky Mtn. J. Math. **7**, No. 3 (1967), 557–567.

[11] S. R. Bernfeld and V. Lakshmikantham, *An Introduction to Nonlinear Boundary Value Problems*, Academic Press, New York, 1974.

[12] D. Bielawski and T. Pruszko, *On the structure of the set of solutions of a functional equation with application to boundary value problems*, Ann. Polon. Math. **53**, No. 3 (1991), 201–209.

[13] P. Binding, *On infinite dimensional differential equations*, J. Diff. Eq. **24** (1977), 349–354.

[14] A. V. Bogatyrev, *Fixed points and properties of solutions of differential inclusions*, Izv. Akad. Nauk SSSR, Ser. Mat. **47**, No. 4 (1983), 895–909.

[15] N. Bourbaki, *Elements de Mathématique XII*, Chap. IV, Hermann, Paris, 1961.

[16] A. Bressan, A. Cellina and A. Fryszkowski, *A class of absolute retracts in spaces of integrable functions*, Proc. Amer. Math. Soc. **112** (1991), 413-418.

[17] F. E. Browder and C. P. Gupta, *Topological degree and nonlinear mappings of analytic type in Banach spaces*, J. Math. Anal. Appl. **26** (1969), 730–738.

[18] A. I. Bulgakov and L. N. Lyapin, *Some properties of the set of solutions of a Volterra-Hammerstein integral inclusion*, Differents. Uravn. **14**, No. 8 (1978), 1043–1048.

[19] A. I. Bulgakov and L. N. Lyapin, *Certain properties of the set of solutions of the Volterra-Hammerstein integral inclusion*, Differents. Uravn. **14**, No. 8 (1978), 1465–1472.

[20] A. I. Bulgakov and L. N. Lyapin, *On the connectedness of sets of solutions of functional inclusions*, Mat. Sbornik **119**, No. 2 (1982), 295–300.

[21] C. Castaing and M. Valadier, *Convex Analysis and Measurable Multifunctions*, Lecture Notes in Math. **580**, Springer-Verlag, Berlin, 1977.

[22] A. Cellina, *On the existence of solutions of ordinary differential equations in Banach space*, Funkc. Ekvac. **14** (1971), 129–136.

[23] A. Cellina, *On the local existence of solutions of ordinary differential equations*, Bull. Acad. Polon. Sci. **20** (1972), 293–296.

[24] A. Cellina, *On the nonexistence of solutions of differential equations in nonreflexive spaces*, Bull. Amer. Math. Soc. **78** (1972), 1069–1072.

[25] J. Chandra, V. Lakshmikantham and A. R. Mitchell, *Existence of solutions of boundary value problems for nonlinear second order systems in a Banach space*, Nonl. Anal. T.M.A. **2** (1978), 157–168.

[26] S. N. Chow and J. D. Schuur, *An existence theorem for ordinary differential equations in Banach spaces*, Bull. Amer. Math. Soc. **77** (1971), 1018–1020.

[27] S. N. Chow and J. D. Schuur, *Fundamental theory of contingent differential equations in Banach spaces*, Trans. Amer. Math. Soc. **179** (1973), 133–144.

[28] A. Constantin, *Stability of solution sets of differential equations with multivalued right hand side*, J. Diff. Eqns. **114** (1994), 243–252.

[29] G. Conti, V.V. Obukhovskii and P. Zecca, *On the topological structure of the solution set for a semilinear functional-differential inclusion in a Banach space*, to appear.

[30] J.-F. Couchouron and M. Kamenskii, *Perturbations d'inclusions paraboliques par des opérateurs condensants*, C. R. Acad. Sci. Paris, **320** (1995), Serie I, 1–6.

[31] J. L. Davy, *Properties of the solution set of a generalized differential equation*, Bull. Austral. Math. Soc. **6**, No. 3 (1972), 379–398.

[32] M. M. Day, *Normed Linear Spaces*, Springer-Verlag, Berlin, 1973.

[33] F. De Blasi, *Existence and stability of solutions for autonomous multivalued differential equations in Banach space*, Rend. Accad. Naz. Lincei, Serie VII, **60** (1976), 767–774.

[34] F. De Blasi, *On a property of the unit sphere in a Banach space*, Bull. Soc. Math. R. S. Roumaine **21** (1977), 259–262.

[35] F. De Blasi, *Characterizations of certain classes of semicontinuous multifunctions by continuous approximations*, J. Math. Anal. Appl. **106**, No. 1 (1985), 1–8.

[36] F. De Blasi and J. Myjak, *On the solutions sets for differential inclusions*, Bull. Polon. Acad. Sci. **33** (1985), 17–23.

[37] F. De Blasi and J. Myjak, *On the structure of the set of solutions of the Darboux problem for hyperbolic equations*, Proc. Edinburgh Math. Soc., Ser. 2, **29**, No. 1 (1986), 7–14.

[38] F. De Blasi and G. Pianigiani, *Solution sets of boundary value problems for nonconvex differential inclusions*, Nonl. Anal. T.M.A. **1** (1993), 303–313.

[39] F. DeBlasi, G. Pianigiani and V. Staicu, *Topological properties of nonconvex differential inclusions of evolution type*, Nonl. Anal. T.M.A. **24** (1995), 711–720.

[40] K. Deimling, *On existence and uniqueness for Cauchy's problem in infinite dimensional Banach spaces*, Proc. Colloq. Math. Soc. Janos Bolyai **15** (1975), 131–142.

[41] K. Deimling, *Ordinary Differential Equations in Banach Spaces*, Lecture Notes in Math. **596**, Springer-Verlag, Berlin, 1977.

[42] K. Deimling, *Periodic solutions of differential equations in Banach spaces*, Man. Math. **24** (1978), 31–44.

[43] K. Deimling, *Open problems for ordinary differential equations in Banach space*, in the book Equazioni Differenziali, Florence, 1978.

[44] K. Deimling, *Nonlinear Functional Analysis,* Springer-Verlag, Berlin, 1985.

[45] K. Deimling and M. R. Mohana Rao, *On solutions sets of multivalued differential equations*, Applicable Analysis, **30** (1988), 129–135.

[46] K. Deimling, *Multivalued Differential Equations*, De Gruyter, New York, 1992.

[47] J. Diestel and J. J. Uhl, *Vector Measures*, Math Surveys **15**, Amer. Math. Soc. Providence, 1977.

[48] J. Dieudonné, *Deux examples singuliers d'équations différentielles*, Acta Sci. Math. (Szeged) **12B** (1950), 38–40.

[49] J. Dubois and P. Morales, *On the Hukuhara-Kneser property for some Cauchy problems in locally convex topological vector spaces*, in Lecture Notes in Math. **964**, 162–170, Springer, Berlin, 1982.

[50] J. Dubois and P. Morales, *Structure de l'ensemble des solutions du probléme de Cauchy sous le conditions de Carathéodory*, Ann. Sci. Math. Quebec **7** (1983), 5–27.

[51] N. Dunford and J. T. Schwartz, *Linear Operators, Part I: General Theory*, Wiley (Interscience), New York, 1958.

[52] V.V. Filippov, *The topological structure of spaces of solutions of ordinary differential equations* (Russian), Uspekhi Mat. Nauk **48** (1993), 103-154.

[53] B.D. Gel'man, *On the structure of the set of solutions for inclusions with multivalued operators*, in Global Analysis - Studies and Applications III, ed. Yu. G. Borisovich and Yu. E. Glikhlikh, Lecture Notes in Math., Vol. **1334**, pp. 60-78, Springer, Berlin, 1988.

[54] A. N. Godunov, *A counterexample to Peano's Theorem in an infinite dimensional Hilbert space*, Vestnik Mosk. Gos. Univ., Ser. Mat. Mek. **5** (1972), 31–34.

[55] A. N. Godunov, *Peano's Theorem in an infinite dimensional Hilbert space is false even in a weakened form*, Math. Notes **15** (1974), 273–279.

[56] A. N. Godunov, *On Peano's Theorem in Banach spaces*, Funct. Anal. Appl. **9**, No. 1 (1975), 53–55.

[57] K. Goebel and W. Rzymowski, *An existence theorem for the equation $\dot{x} = f(t, x)$ in Banach spaces*, Bull. Acad. Polon. Math. **18** (1970), 367–370.

[58] L. Gorniewicz and T. Pruszko, *On the set of solutions of the Darboux problem for some hyperbolic equations*, Bull. Acad. Polon. Math. **28**, No. 5-6 (1980), 279–286.

[59] L. Gorniewicz, J. Bryszewski and T. Pruszko, *An application of the topological degree theory to the study of the Darboux problem for hyperbolic equations*, J. Math. Anal. Appl. **76** (1980), 107–115.

[60] L. Gorniewicz, *On the solution sets of differential inclusions*, J. Math. Anal. Appl. **113** (1986), 235–244.

[61] C. P. Gupta, J. J. Nieto and L. Sanchez, *Periodic solutions of some Lienard and Duffing equations*, J. Math. Anal. Appl. **140** (1989), 67–82.

[62] G. Haddad, *Topological properties of the sets of solutions for functional differential inclusions*, Nonl. Anal. T.M.A. **5**, No. 12 (1981), 1349–1366.

[63] A.J. Heunis, *Continuous dependence of the solutions of an ordinary differential equation*, J. Diff. Eqns. **54** (1984), 121-138.

[64] C. Himmelberg and F. Van Vleck, *A note on the solution sets of differential inclusions*, Rocky Mtn. J. Math. **12** (1982), 621–625.

[65] E. Horst, *Differential equations in Banach space: five examples*, Arch. Math. **46** (1986), 440–444.

[66] J. Horváth, *Topological Vector Spaces and Distributions*, Addison-Wesley, Reading, 1966.

[67] T. S. Hu, *Cohomology Theory*, Markham, Chicago, 1968.

[68] M. Hukuhara, *Sur les systèmes des équations differentielles ordinaires*, Jap. J. Math. **5** (1928), 345–350.

[69] D. M. Hyman, *On decreasing sequences of compact absolute retracts*, Fund. Math. **64** (1969), 91–97.

[70] J. Jarnik and J. Kurzweil, *On conditions on right hand sides of differential relations*, Casopis pro Pest. Mat. **102** (1977), 334–349.

[71] M. I. Kamenskii, *On the Peano Theorem in infinite dimensional spaces*, Mat. Zametki **11**, No. 5 (1972), 569–576.

[72] Z. Kánnai and P. Tallos, *Stability of solution sets of differential inclusions*, to appear.

[73] R. Kannan, J. J. Nieto and M. B. Ray, *A class of nonlinear boundary value problems without Landesman-Lazer condition*, J. Math. Anal. Appl. **105** (1985), 1–11.

[74] A. Kari, *On Peano's Theorem in locally convex spaces*, Studia Math. **73**, No. 3 (1982), 213–223.

[75] J. L. Kelley and I. Namioka, *Linear Topological Spaces*, Van Nostrand, Princeton, 1963.

[76] W. G. Kelley, *A Kneser theorem for Volterra integral equations*, Proc. Amer. Math. Soc. **40**, No. 1 (1973), 183–190.

[77] M. Kisielewicz, *Multivalued differential equations in separable Banach spaces*, J. Optim. Th. Appl. **37**, No. 2 (1982), 231–249.

[78] H. Kneser, *Uber die Lösungen eine system gewöhnlicher differential Gleichungen, das der lipschitzschen Bedingung nicht genügt*, S. B. Preuss. Akad. Wiss. Phys. Math. Kl. **4** (1923), 171–174.

[79] A. Kolmogorov and S. Fomin, *Elements of the Theory of Functions and Functional Analysis*, Graylock, New York, 1957.

[80] M. A. Krasnoselskii and S. G. Krein, *Theory of ordinary differential equations in a Banach space*, Trud. Sem. Funkts. Anal. Voronezhsk. Gos. Univ. **2** (1956), 3–23.

[81] P. Krbec and J. Kurzweil, *Kneser's theorem for multivalued differential delay equations*, Casopis pro Pest. Mat. **104**, No. 1 (1979), 1–8.

[82] I. Kubiaczyk and S. Szufla, *Kneser's Theorem for weak solutions of ordinary differential equations in Banach spaces*, Publ. Inst. Math. (Beograd) (NS), **32**, No. 46 (1982), 99–103.

[83] I. Kubiaczyk, *Kneser's Theorem for differential equations in Banach spaces*, J. Diff. Eq. **45**, No. 2 (1982), 139–147.

[84] I. Kubiaczyk, *Structure of the sets of weak solutions of an ordinary differential equation in a Banach space*, Ann. Polon. Math. **44**, No. 1 (1980), 67–72.

[85] V. Lakshmikantham and G. E. Ladas, *Differential Equations in Abstract Spaces*, Academic Press, New York, 1972.

[86] A. Lasota and J. A. Yorke, *The generic property of existence of solutions of differential equations in Banach spaces*, J. Diff. Eqns. **13** (1973), 1–12.

[87] J. M. Lasry and R. Robert, *Analyse Non Linéaire Multivoque*, Cahiers de Math. de la Decision, Paris, No. 7611, 1977.

[88] J. M. Lasry and R. Robert, *Acyclicité de l'ensemble des solutions de certains equations functionelles*, C. R. Acad. Sci. Paris, **282**, No. 22A (1976), 1283–1286.

[89] T. C. Lim, *On fixed point stability for set valued contractive mappings with applications to generalized differential equations*, J. Math. Anal. Appl. **110** (1985), 436–441.

[90] A. Margheri and P. Zecca, *A note on the topological structure of solution sets of Sturm-Liouville problems in Banach spaces*, Atti Accad. Naz. Lincei, Rend. Cl. Sci. Fis. Math. Nat., to appear.

[91] H. Monch, *Boundary value problems for nonlinear ordinary differential equations of second order in Banach spaces*, Nonl. Anal. T.M.A. **4** (1980), 985–999.

[92] H. Monch and G. von Harten, *On the Cauchy problem for ordinary differential equations in Banach spaces*, Arch. Math. **39** (1982), 153–160.

[93] A. M. Muhsinov, *On differential inclusions in Banach space*, Soviet Math. Dokl. **15** (1974), 1122–1125.

[94] J. J. Nieto, *Periodic solutions of nonlinear parabolic equations*, J. Diff. Eqns. **60**, No. 1 (1985), 90–102.

[95] J. J. Nieto, *Structure of the solution set for semilinear elliptic equations*, Colloq. Math. Soc. Janos Bolyai, **47** (1987), 799–807.

[96] J. J. Nieto, *Nonuniqueness of solutions of semilinear elliptic equations at resonance*, Boll. U.M.I. **6, 5–A**, No. 2, (1986), 205–210.

[97] J. J. Nieto, *Hukuhara-Kneser property for a nonlinear Dirichelet problem*, J. Math. Anal. Appl. **128** (1987), 57–63.

[98] J. J. Nieto, *Decreasing sequences of compact absolute retracts and nonlinear problems*, Boll. Un. Mat. Ital. **2-B**, No. 7 (1988), 497–507.

[99] J. J. Nieto, *Aronszajn's theorem for some nonlinear Dirichelet problem*, Proc. Edinburg Math. Soc. **31** (1988), 345–351.

[100] J. J. Nieto and L. Sanchez, *Periodic boundary value problems for some Duffing equations*, Diff. and Int. Eq. **1**, No. 4 (1988), 399–408.

[101] J. J. Nieto, *Nonlinear second order periodic value problems with Caratheodory functions*, Appl. Anal. **34** (1989), 111–128.

[102] J. J. Nieto, *Periodic Neumann boundary value problem for nonlinear parabolic equations and application to an elliptic equation*, Ann. Polon. Math. **54**, No. 2 (1991), 111–116.

[103] V.V. Obukhovskii, *Semilinear funciional differential inclusions in a Banach space and controlled parabolic systems*, Soviet J. Automat. Inform. Sci. **24**, No. 3 (1991), 71–79.

[104] C. Olech, *On the existence and uniqueness of solutions of an ordinary differential equation in the case of Banach space*, Bull. Acad. Polon. Math. **8** (1969), 667–673.

[105] N. S. Papageorgiou, *Kneser's Theorem for differential equations in Banach spaces*, Bull. Austral. Math. Soc. **33**, No. 3 (1986), 419–434.

[106] N. S. Papageorgiou, *On the solution set of differential inclusions in Banach space*, Appl. Anal. **25**, No. 4 (1987), 319–329.

[107] N. S. Papageorgiou, *A property of the solution set of differential inclusions in Banach spaces with a Caratheodory orientor field*, Appl. Anal. **27**, No. 4 (1988), 279–287.

[108] N. S. Papageorgiou, *On the solution set of differential inclusions with state constraints*, Appl. Anal. **31** (1989), 279–289.

[109] N. H. Pavel and J. Vrabie, *Equations d'evolution multivoques dans le espace de Banach*, Comptes Rendus Acad. Sci. Paris, Serie A, **287** (1978), 315–317.

[110] G. Peano, *Sull'integrabilità delle equazioni differenziali del primo ordine*, Atti della Reale Accad. dell Scienze di Torino **21** (1886), 677–685.

[111] G. Peano, *Démonstration de l'integrabilité des equations differentielles ordinaires*, Mat. Annalen **37** (1890), 182–238.

[112] A. Peixoto and R. Thom, *Le point de vue enumeratif dans les problemes aux limites pour les equations differentielles ordinaires II. Le theoremes*, Comptes Rendus Acad. Sci. Paris, Ser. 1 Math., **303** (1968), 693-698; erratum **307** (1988), 197-198.

[113] G. Pianigiani, *Existence of solutions of ordinary differential equations in Banach spaces*, Bull. Acad. Polon. Math. **23** (1975), 853–857.

[114] S. Plaskacz, *On the solution sets for differential inclusions*, Boll. U.M.I. **7, 6-A** (1992), 387–394.

[115] B. Ricceri, *Une propriété topologique de l'ensemble des points fixed d'une contraction multivoque à valeurs convexes*, Atti Accad. Naz. Lincei Cl. Sci. Fis. Mat. Natur. **51**(5), (1987), 283–286.

[116] B. Ricceri, *The existence of solutions for ordinary differential equations in Banach spaces under Carathéodory hypotheses*, Boll. Un. Mat. Ital. C(5), **18** (1981), 1–19.

[117] L. Rybinski, *On Caratheodory type selections*, Fund. Math. **125**, (1985), 187–193.

[118] B. Rzepecki, *On measures of noncompactness in topological vector spaces*, Comment. Math. Univ. Carolinae, **23**, No. 1 (1982), 105–116.

[119] B. Rzepecki, *On the equation $\dot{y} = f(t,y)$ in Banach spaces*, Comment. Math. Univ. Carolinae, **24**, No. 4 (1983), 609–620.

[120] B. Rzepecki, *Note on differential equations in Banach spaces*, Rev. Roum. Math. Pures Appl. **30**, No. 8 (1985), 679–684.

[121] T. Rzeżuchowski, *Scorza-Dragoni type theorems for upper semicontinuous multivalued functions*, Bull. Acad. Polon. Sci., ser. sci. math., **28** (1980), 61–67.

[122] B. N. Sadovskii, *On measures of noncompactness and contracting operators*, in Problems in the Mathematical Analysis of Complex Systems (Russian), second edition, Voronezh (1968), 89–119.

[123] B. N. Sadovskii, *Limit-compact and condensing operators* (in Russian), Uspekh. Mat. Nauk **27** (1972), 1–146.

[124] K. Schmitt and P. Volkmann, *Boundary value problems for second order differential equations in convex subsets in a Banach space*, Trans. Amer. Math. Soc. **218** (1976), 397–405.

[125] R. M. Sentis, *Convergence de solutions d'equations differentielles multivoques*, Comptes Rendus Acad. Sci. Paris, Serie A, **278** (1974), 1623–1626.

[126] J. S. Shin, *Kneser type theorems for functional differential equations in a Banach space*, Funk. Ekvacioj **35** (1992), 451–466.

[127] Z. Song, *Existence of generalized solutions for ordinary differential equations in Banach spaces*, J. Math. Anal. Appl. **128** (1987), 405–412.

[128] W. Sosulski, *Compactness and upper semicontinuity of solution set of functional differential equations of hyperbolic type*, Comment. Mat. Prace. Mat. **25**, No. 2 (1985), 359–362.

[129] G. Stampacchia, *Le trasformazioni che presentano il fenomeno di Peano*, Rend. Accad. Naz. Lincei, **7** (1949), 80–84.

[130] A. Szép, *Existence theorem for weak solutions of ordinary differential equations in reflexive Banach spaces*, Studia Sci. Math. Hungar. **6** (1971), 197–203.

[131] S. Szufla, *Some remarks on ordinary differential equations in Banach spaces*, Bull. Acad. Polon. Math. **16** (1968), 795–800.

[132] S. Szufla, *Measure of noncompactness and ordinary differential equations in Banach spaces*, Bull Acad. Polon. Sci. **19** (1971), 831–835.

[133] S. Szufla, *Structure of the solutions set of ordinary differential equations in Banach space*, Bull. Acad. Polon. Sci. **21**, No. 2 (1973), 141–144.

[134] S. Szufla, *Solutions sets of nonlinear equations*, Bull. Acad. Polon. Sci. **21**, No. 21 (1973), 971–976.

[135] S. Szufla, *Some properties of the solutions set of ordinary differential equations*, Bull. Acad. Polon. Sci. **22**, No. 7 (1974), 675–678.

[136] S. Szufla, *On the structure of solutions sets of differential and integral equations in Banach spaces*, Ann. Polon. Math. **34** (1977), 165–177.

[137] S. Szufla, *On the equation $\dot{x} = f(t,x)$ in Banach spaces*, Bull. Acad. Polon. Sci. **26**, No. 5 (1978), 401–406.

[138] S. Szufla, *Kneser's Theorem for weak solutions of ordinary differential equations in reflexive Banach spaces*, Bull. Acad. Polon. Sci. **26**, No. 5 (1978), 407–413.

[139] S. Szufla, *On the existence of solutions of differential equations in Banach spaces*, Bull. Acad. Polon. Sci. **30**, No. 11–12 (1982), 507–515.

[140] S. Szufla, *On the equation $\dot{x} = f(t,x)$ in locally convex spaces*, Math. Nachr. **118** (1984), 179–185.

[141] S. Szufla, *Existence theorems for solutions of integral equations in Banach spaces*, Proc. Conf. Diff. Eq. and Optimal Control, Zielona Gora (1985), 101–107.

[142] S. Szufla, *On the application of measure of noncompactness to differential and integral equations in Banach space*, Fasc. Math. **18** (1988), 5–11.

[143] S. Szufla, *On the structure of solution sets of nonlinear equations*, Proc. Conf. Diff. Eq. and Optimal Control, Zielona Gora (1989), 33-39.

[144] S. Szufla and D. Bugajewski, *Kneser's theorem for weak solutions of the Darboux problem in Banach spaces*, Nonl. Anal. T.M.A. **20**, No. 2 (1993), 169–173.

[145] P. Talaga, *The Hukuhara-Kneser Property for parabolic systems with nonlinear boundary conditions*, J. Math. Anal. **79** (1981), 461–488.

[146] P. Talaga, *The Hukuhara-Kneser property for quasilinear parabolic equations*, Nonl. Anal. T.M.A. **12**, No. 3 (1988), 231–245.

[147] A. A. Tolstonogov, *On differential inclusions in Banach space and continuous selectors*, Dokl. Akad. Nauk SSSR, **244** (1979), 1088–1092.

[148] A. A. Tolstonogov, *On properties of solutions of differential inclusions in Banach space*, Dokl. Akad. Nauk SSSR, **248** (1979), 42–46.

[149] A. A. Tolstonogov, *On the structure of the solution set for differential inclusions in a Banach space*, Math. Sbornik, **46** (1983), 1–15.

[150] A. A. Tolstonogov, *Differential Inclusions in Banach Spaces*, Izdat. Nauka, Novosibirsk, 1986 (in Russian).

[151] G. Vidossich, *On Peano phenomenon*, Bull. Un. Math. Ital. **3** (1970), 33–42.

[152] G. Vidossich, *On the structure of the set of solutions of nonlinear equations*, J. Math. Anal. Appl. **34** (1971), 602–617.

[153] G. Vidossich, *A fixed point theorem for function spaces*, J. Math. Anal. Appl. **36** (1971), 581–587.

[154] G. Vidossich, *Existence, uniqueness and approximation of fixed points as a generic property*, Bol. Soc. Brasil. Mat., **5** (1974), 17–29.

[155] G. Vidossich, *Two remarks on global solutions of ordinary differential equations in the real line*, Proc. Amer. Math. Soc. **55** (1976), 111–115.

[156] T. Wazewski, *Sur l'existence et l'unicité des integrales des équations différentielles ordinaires au cas de l'espace de Banach*, Bull. Acad. Polon. Math. **8** (1960), 301–305.

[157] J. A. Yorke, *Spaces of solutions*, Lect. Notes Op. Res. Math. Econ. **12**, Springer-Verlag, (1969), 383–403.

[158] J. A. Yorke, *A continuous differential equation in Hilbert space without existence*, Funkc. Ekvac. **13** (1970), 19–21.

Roberto Dragoni
Dipartimento di Matematica,
Università di Siena,
Siena, Italy

Jack W. Macki
Department of Mathematics,
University of Alberta,
Edmonton, Alberta, Canada   T6G 2G1

Paolo Nistri and Pietro Zecca
Dipartimento di Sistemi e Informatica,
Facoltà di Ingegneria,
Università di Firenze,
via S. Marta 3,
50139 Firenze, Italy.